SCIENCE PLUS®
TECHNOLOGY AND SOCIETY

LEVEL BLUE

TEACHING RESOURCES

Unit 5
Electromagnetic Systems

HOLT, RINEHART AND WINSTON
Harcourt Brace & Company
Austin • New York • Orlando • Atlanta • San Francisco • Boston • Dallas • Toronto • London

To the Teacher

This booklet contains a comprehensive collection of teaching resources. You will find all of the blackline masters that you need to plan, implement, and assess this unit. Also included are worksheets that correspond directly to the SourceBook.

Choose from the following blackline masters to meet your needs and the needs of your students:

- **Home Connection** consists of a parent letter designed to get parents involved in the excitement of the *SciencePlus* method. The letter provides parents with a general idea of what you are going to cover in the unit, and it even gives you an opportunity to ask for any common household materials that you may need to accomplish the unit's activities most economically.

- **Discrepant Event Worksheets** provide demonstrations and activities that seem to challenge logic and reason. These worksheets motivate students to question their previous knowledge and to develop reasonable explanations for the discrepant phenomena.

- **Math Practice Worksheets** and **Graphing Practice Worksheets** help fine-tune math and graphing skills.

- **Theme Worksheets** encourage students to make connections among the major science disciplines.

- **Spanish Resources** include Spanish versions of the Home Connection letter, plus worksheets that outline the big ideas and principal vocabulary terms for the unit.

- **Transparency Worksheets** correspond to teaching transparencies to help you reteach, extend, or review major concepts.

- **SourceBook Activity Worksheets** reinforce content introduced in the SourceBook.

- **Resource Worksheets** consist of blackline-master versions of charts, graphs, and activities in the Pupil's Edition.

- **Exploration Worksheets** consist of blackline-master versions of Explorations in the Pupil's Edition. To help students focus on specific tasks, many of these worksheets include a goal, step-by-step instructions, and even cooperative-learning strategies. These worksheets simplify the tasks of assigning homework, allowing opportunities for make-up work, and providing lesson plans for substitute teachers.

- **Unit Activity Worksheet** consists of an activity, such as a crossword puzzle or word search, that provides a fun way for students to review vocabulary and main concepts.

- **Review Worksheets** consist of blackline-master versions of the review materials in the Pupil's Edition, including Challenge Your Thinking, Making Connections, and SourceBook Unit CheckUp.

- **Chapter Assessments** and **End-of-Unit Assessments** provide additional assessment questions. Each assessment worksheet includes two or more Challenge questions that encourage students to synthesize the main concepts of the chapter or unit and to apply them in their own lives.

- **Activity Assessments** are activity-based assessment worksheets that allow you to evaluate students' ability to solve problems using the tools, equipment, and techniques of science.

- **Self-Evaluation of Achievement** gives you an easy method of monitoring student progress by allowing students to evaluate themselves.

- **SourceBook Assessment** is an easy-to-grade test consisting of multiple-choice, true-false, and short-answer questions.

For your convenience, an **Answer Key** is available in the back of this booklet. The key includes reduced versions of all applicable worksheets, with answers included on each page.

Credits: See page 89.

SCIENCEPLUS is a registered trademark of Harcourt Brace & Company licensed to Holt, Rinehart and Winston, Inc.

Printed in the United States of America

ISBN 0-03-095693-5   1 2 3 4 5 6 7 8 9 021 99 98 97 96

Unit 5: Electromagnetic Systems

Contents

▼ *A corresponding transparency is available. See the Teaching Transparencies Cross-Reference on the next page.*

Contents, continued

Teaching Transparencies Cross-Reference

Dear Parent,

In the next few weeks, your son or daughter will be learning more about electricity. Different sources of electricity will be investigated, as well as how electrical currents flow through circuits. The relationship between electricity and magnetism will also be explored. By the time the students have finished Unit 5, they should be able to answer the following questions to grasp the "big ideas" of the unit.

1. How may electricity be produced? (Ch. 13)

2. What is the difference between a cell and a battery? (Ch. 13)

3. What parts are common to all chemical cells, and how do they operate? (Ch. 13)

4. How do AC and DC currents differ? (Ch. 14)

5. What factors affect the size of currents in circuits? (Ch. 15)

6. What are some differences between series and parallel circuits? (Ch. 15)

7. What are electromagnets? How can their strength be increased? (Ch. 15)

8. How is the flow of water like that of electricity? (Ch. 15)

9. By what units is electricity measured? How are these units related? (Ch. 15)

Here is an activity that you may want to do at home. Have your son or daughter take a compass around the house or apartment and hold it near various appliances. Have him or her note any movement of the compass needle at various positions around each appliance. Your son or daughter should check both appliances that are turned on and appliances that are turned off. Have him or her move the compass as close to the motors as is safely possible. (The compass needle should be deflected when it is brought close to an operating electric motor.) Ask your son or daughter to analyze his or her findings. Discuss the possible causes. (Basically, the needle will be deflected near an operating motor because the motor creates a magnetic field as electricity is converted into motion.) This simple activity will help students see the relationship between electricity and magnetism. Students will learn much more about the relationship between electricity and magnetism as they progress through the unit.

Sincerely,

The items listed below are materials that we will use in class for the various science explorations of Unit 5. Your contribution of materials would be very much appreciated. I have checked certain items below. If you have these items and are willing to donate them, please send them to the school with your son or daughter by

_____.

- ○ aluminum foil
- ○ cardboard (sheets and tubes)
- ○ clothes hangers (wire)
- ○ clothespins
- ○ compasses (magnetic)
- ○ copper strips
- ○ corks
- ○ cups (small; paper)
- ○ D-cell batteries
- ○ flannel cloth
- ○ flashlight bulbs
- ○ fluorescent or neon lights
- ○ forceps
- ○ iron bolts
- ○ knitting needles (short)
- ○ latex gloves
- ○ light bulbs
- ○ magnets
- ○ masking tape
- ○ modeling clay
- ○ newspapers

- ○ paper clips
- ○ plastic strips or plastic rulers
- ○ plastic wrap
- ○ rubber bands
- ○ salt
- ○ sandpaper
- ○ screws
- ○ solar cells
- ○ steel wool
- ○ straight pins
- ○ thread
- ○ thumbtacks
- ○ vinyl strips, about 10 cm long
- ○ washers
- ○ wheat puffs
- ○ wire (copper, iron, and nichrome— insulated and uninsulated)
- ○ wooden blocks (small)
- ○ wooden dowels
- ○ zinc strips

Thank you in advance for your help.

EXPLORATION 1

Electricity Working for You, page 296

Your goal	to use electricity to produce other forms of energy for use	**Safety Alert!**

You have probably seen warnings like the one at right on appliances or power tools. As useful as it is, electricity in large amounts is deadly and must be carefully controlled.

The familiar situations that you have been analyzing in the Case Studies on pages 294 and 295 of your textbook involve complex electrical parts and arrangements. Large quantities of electricity are used in the devices in Case Studies A and B. This is true for the operation of most appliances in homes, stores, or industry. A moderate amount is used in the devices in Case Studies C and D. Care must be exercised in using these amounts of electricity.

Most of the Explorations in this unit require only a small amount

Illustration also on page 296 of your textbook

of electricity. They are quite safe to do. Here are four experiments in which a small amount of electricity works for you by producing other forms of energy. Watch for these energy forms.

In these experiments the electricity is supplied by dry cells such as those used in flashlights.

Activity 1: Shedding a Little Light

You Will Need

- a length of magnet wire
- sandpaper
- a flashlight bulb
- a D-cell
- a rubber band

What to Do

1. Begin by sanding the enamel off the last few centimeters of each end of the wire to expose the copper underneath. Then, using only the other items listed, find all of the different arrangements that will light the bulb.

2. In the space below, sketch each arrangement.

Name _____ Date _____ Class _____

3. What form(s) of energy does the electricity produce?

_____

4. What are other examples of electricity being used in this way?

_____

_____

5. You have constructed an electric *circuit.* What parts make up this circuit?

_____

_____

Check the dictionary to find out the origin of the word *circuit.* How is it significant?

_____

_____

_____

What would you say an electric circuit is?

_____

_____

_____

_____

Activity 2: The Heat Is On

You Will Need

- two 30 cm lengths of magnet wire
- sandpaper
- modeling clay
- a clothespin
- 2 D-cells
- a wide rubber band
- a strand of steel wool
- aluminum foil
- a thin nichrome wire (10 cm long)

What to Do

1. Sand the enamel off of the last 5 cm of the ends of each length of magnet wire. Then make a small loop at one end of each wire.

2. Bend the wires and support them with modeling clay, as shown on page 297 of your textbook.

3. Attach the ends of the wires to the D-cell, securing them with a wide rubber band.

4. Place a strand of steel wool through the loops and let it rest in contact with each loop. What do you observe? **(NOTE: Do not leave the test materials in contact with the loops for very long because this will quickly drain the cell of its electrical energy.)**

_____

_____

Exploration 1 Worksheet, continued

Repeat with a rolled-up length of aluminum foil and then with a length of nichrome wire. Record your observations here.

_____

_____

_____

Bring your hand close to each piece being tested, but do not touch any of them. What do you feel?

_____

_____

5. What form of energy does the electricity produce?

_____

6. What are some examples of devices in which this type of electrical energy is used?

_____

_____

_____

7. Do Activities 1 and 2 demonstrate the same principle? Explain.

_____

_____

_____

_____

_____

_____

As you may have guessed from the title of this unit, there is a connection between electricity and magnetism. The following Activities will help illustrate that connection.

Activity 3: The Electricity-Magnetism Connection

You Will Need

- a D-cell
- a compass
- 2 thumbtacks or screws
- a wood block
- a paper clip
- 2 lengths of magnet wire (15 cm and 25 cm long)
- a rubber band
- sandpaper

What to Do

1. Sand the enamel off of the last 2 or 3 cm of the ends of each wire. Then set up the apparatus as shown below. Align the compass needle, and place the wire over the compass in a north-south direction so that the wire lines up with the compass needle.

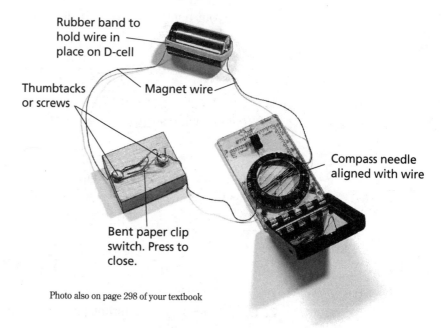

Rubber band to hold wire in place on D-cell

Thumbtacks or screws

Magnet wire

Compass needle aligned with wire

Bent paper clip switch. Press to close.

Photo also on page 298 of your textbook

2. Close the electrical circuit by pressing the contact switch. What happens? **Caution: Don't keep the switch closed for very long.**

_____

_____

_____

3. What kind of energy does the electricity produce in this experiment?

_____

4. Can you think of any everyday applications that make use of electricity in this way?

_____

_____

_____

_____

_____

_____

Exploration 1 Worksheet, continued

_____

Activity 4: Let's Get Moving!

You Will Need

- 2 D-cells
- a paper clip
- 2 lengths of magnet wire (10 cm and 250 cm long)
- a bar magnet
- a support stand with ring clamp
- a cork
- a wide rubber band
- a narrow strip of cardboard (2 cm × 30 cm)
- 2 thumbtacks or screws
- a wood block
- sandpaper

What to Do

1. Sand the enamel off of the last 2 or 3 cm of the ends of each wire. Then assemble the circuit as shown on page 299 of your textbook.

2. Have one person hold a strong magnet near the cork while the other person presses the contact switch to complete the circuit. Observe what happens. **Caution: Don't keep the switch closed for very long.**

_____

_____

Open the switch. What happens?

_____

_____

_____

3. Using the other end of the magnet, repeat step 2. What happens?

_____

_____

_____

_____

_____

_____

4. What kind of energy is produced by the electricity?

_____

_____

5. What are some practical examples of how energy works for you in this way?

_____

_____

A Home Project

Make a battery tester. Use one of the arrangements you discovered in Activity 1 of Exploration 1. Devise a tester that consists of a light bulb with two wires connected to it. Touch the ends of the wire to the battery. The brightness of the light bulb will indicate the strength of the battery.

Bending Water   Teacher Demonstration

Conduct this activity before beginning Lesson 2, What Is Electricity? which begins on page 300 of the textbook, as an introduction to charge, electrons, and static electricity.

You Will Need

- a water faucet
- a plastic comb
- a piece of wool cloth

What to Do

1. Turn on the water faucet so that the water runs in a steady but slow, thin stream.

2. Ask the students: How could I make the stream of water bend toward the comb without touching the water? *(Accept all reasonable responses, but do not provide students with the answer.)*

3. Rub the comb vigorously with the wool cloth, and hold the comb near the stream of water. The stream of water should bend toward the comb.

4. Ask: What made this happen? *(Students should answer that something about rubbing the comb with the wool caused the water to be attracted to the comb. Some students may even respond that it was caused by static electricity. If so, ask them to try to explain how static electricity caused this to happen.)*

Explanation

When the comb was rubbed with the wool, electrons were rubbed onto the comb. This caused the comb to become negatively charged. This buildup of charges is known as *static electricity*. The water molecules have both positive and negative charges. The positive ends of the water molecules were attracted to the negatively charged electrons, causing the stream of water to move toward the comb. Assure students that although many of these terms may seem unfamiliar to them now, they will become more familiar with them as they work through this lesson.

Discussion

Generate further class discussion with questions such as the following:

- What properties must electrons have if they can be rubbed off so easily? *(Expected answer: They must be very light and able to move about easily.)*

- Do you know of any other examples of static electricity? *(Possible responses: clothes clinging together after coming out of the dryer, getting a shock when touching a metal doorknob, hair sticking up after putting on a sweater)*

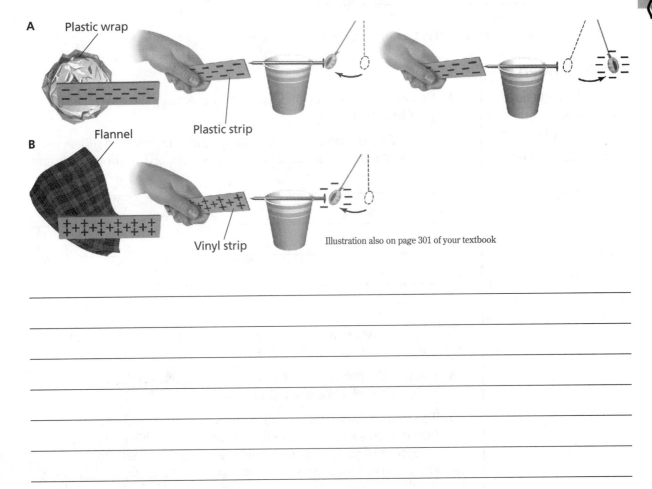

Name _____ Date _____ Class _____

A Theory of Charged Particles, continued

You might experiment by replacing the nail in the demonstration with a glass rod, a wooden stick, a copper wire, or objects made of other materials. Which are conductors? Which are insulators?

_____

_____

The drawings below apply the theory just stated to the previous demonstration. Express in your own words what is taking place in each drawing.

A Plastic wrap

Plastic strip

B Flannel

Vinyl strip

Illustration also on page 301 of your textbook

_____

_____

_____

_____

_____

_____

Name ______________________ Date ____________ Class ____________

EXPLORATION 2

Chapter 13
Exploration Worksheet

Constructing a Current Detector, page 306

Your goal	to build a galvanometer, an instrument used to detect and measure electrical current

Your task is to design and construct a homemade galvanometer.

You Will Need

- a small magnetic compass
- insulated wire
- anything else you need to hold the parts in place
- paper clip
- magnet wire
- sandpaper
- paper cup
- concentrated lemon juice

Hints

1. Remember the results of Exploration 1, Activity 3 on page 298 of your textbook.
2. Try coiling the insulated wire around the compass. Use different numbers of turns and observe the effect. Record your observations.

 ____

 ____

 ____

 ____

 ____

3. Leave the two ends of the wire free so that you can attach them to the source of the small current.
4. Position the galvanometer so that the compass needle and the coil of wire are parallel to one another.
5. Test your galvanometer using a small current, which can be obtained from a lemon-juice cell. To make a lemon-juice cell, place a straightened paper clip and a piece of sanded magnet wire into a small cup of concentrated lemon juice. The paper clip and the wire are the electrodes, and the lemon juice is the electrolyte. Hook the free ends of the galvanometer wire to the electrodes. What happens?

 ____

 ____

6. Will your galvanometer be able to give you any information about the size of the current? Explain.

 ____

 ____

 ____

Challenge Your Thinking, page 307

1. All Charged Up

a. One by one, a negatively charged plastic ruler is brought near three light, foil-covered spheres suspended by nylon threads. The ruler repels sphere *A* and attracts spheres *B* and *C*. Sphere *A* attracts sphere *B*, and sphere *C* attracts sphere *B*. Do you have enough information to determine the charges on spheres *B* and *C*? Why or why not?

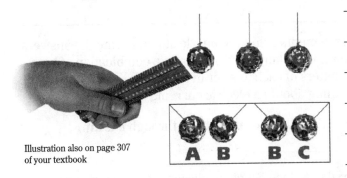

Illustration also on page 307 of your textbook

b. Jeff rubbed two pieces of plastic wrap with a sock and then suspended them, as shown on page 307 of your textbook. Note what he observed.

c. Jeff then brought one of the pieces near a table. Again observe what he saw.

d. Explain what is happening.

Try Jeff's experiment yourself.

2. Sure Shot

When spray painting a screen, the screen is given an electric charge. Why doesn't the paint spray go through or around the screen?

Look, no paint!

Illustration also on page 307 of your textbook

_____

_____

_____

_____

_____

_____

3. Current Puzzle

Complete the following statements using the puzzle at left. Locate the answers in the puzzle by searching horizontally, vertically, or in a combination of both directions. Cross out the letters of each answer in the puzzle. The letters that remain will tell you something about an electric current.

E	N	N	O	R	T	C	U
G	R	E	L	E	C	R	E
A	T	I	V	E	N	T	R
T	C	U	D	N	O	C	O
O	I	N	S	U	L	A	T
R	=	C	H	I	T	I	V
P	A	R	G	S	U	E	E
R	O	T	O	O	N	D	E
S	I	N	N	P	C	M	G
O	T	I	O	N	H	A	R

a. A material that allows charges to go through it is a(n) _____.

b. A material that does not allow charges to go through it is a(n) _____.

c. J. J. Thomson discovered a charged particle that moves readily; it is called a(n) _____.

d. This type of particle has a(n) _____ charge.

e. The other charged particle in materials, which is more massive and does not move, is called a(n) _____ and has a(n) _____ charge.

f. If a material has equal numbers of these two kinds of particles, the material is _____.

Mystery statement: _____

4. That's a Wrap

Some kinds of plastic wrap can be stretched tightly over a container and down its sides. The plastic wrap sticks to the sides of the container. Why?

_____

_____

_____

_____

Name _____ Date _____ Class _____

Word Usage

1. Write one or two sentences to explain the difference between the words *electrolyte* and *electrode*.

_____

_____

_____

_____

_____

Correction/ Completion

2. The following sentences are incorrect or incomplete. Your challenge is to make them correct and complete.

a. When one material is rubbed against another, electricity causes charged particles to move from one material to the other.

_____

_____

_____

b. An electrostatic charge involves the continuous flow of electrons through a material such as a copper wire.

_____

_____

_____

Short Responses

3. Identify each of these setups with a single term.

a. A free-floating magnetic needle used to tell direction _____

b. A group of connected cells _____

c. A wire coiled around a compass to indicate an electric current _____

4. Circle the materials that are good conductors of electricity.

    a. glass   **b.** wood

    c. iron   **d.** zinc

    e. acid   **f.** plastic

Illustration for Interpretation

5. Use what you've learned about static electricity to explain the illustration below.

_____

_____

_____

_____

Short Response

6. Name at least one way your life would be different if plastic were not a good insulator.

_____

_____

_____

_____

EXPLORATION 1

Chemical Cells, page 310

Your goal	to make and test a wet cell and a dry cell	**Safety Alert!**
		Wear goggles and latex gloves when working with ammonium chloride. 

Experiment 1: Dry Cells

You Will Need

- a homemade or commercial galvanometer
- a 40 cm length of magnet wire
- masking tape
- rubber bands
- 2 zinc strips (3 cm × 8 cm)
- 2 copper strips (3 cm × 8 cm)
- blotting paper or filter paper
- ammonium chloride solution
- a container to hold the ammonium chloride solution
- forceps
- latex gloves

What to Do

1. Make a single-cell sandwich like the one shown below at left.

2. Measure the amount of deflection this sandwich produces in your galvanometer.

3. Now make a double-cell sandwich, as shown in the second illustration below.

4. Measure the deflection it produces in the galvanometer. How does it compare with the deflection caused by the single-cell sandwich?

_____

_____

Because the double-cell sandwich consists of more than a single cell, it is known as a **battery**.

Single-Cell Sandwich

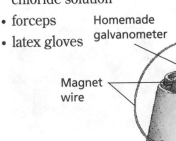

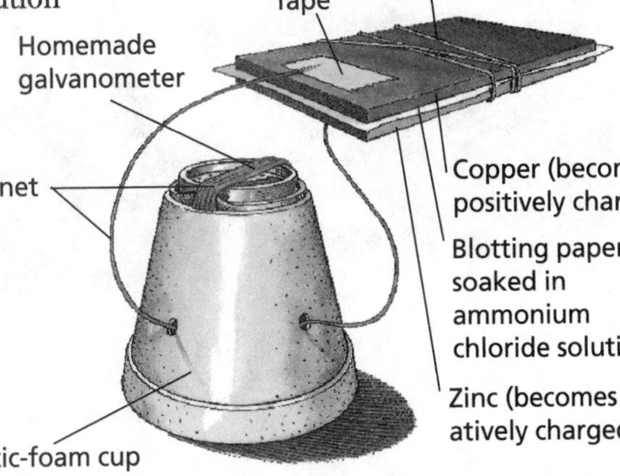

Rubber bands
Tape
Homemade galvanometer
Magnet wire
Copper (becomes positively charged)
Blotting paper soaked in ammonium chloride solution
Zinc (becomes negatively charged)
Plastic-foam cup

Double-Cell Sandwich

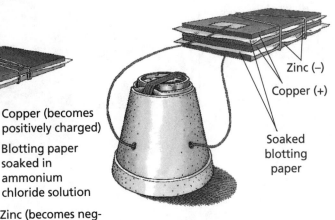

Zinc (−)
Copper (+)
Soaked blotting paper

Illustrations also on page 310 of your textbook

_____

Questions

1. What accounts for the difference in the galvanometer readings for the single-cell and double-cell sandwiches?

_____

_____

What would be the effect of adding more layers to the sandwich?

_____

_____

2. Which electrodes were linked together in converting a single-cell sandwich to a double-cell sandwich?

_____

_____

_____

Experiment 2: Wet Cells

You Will Need

- a commercial galvanometer
- a zinc strip (2 cm × 15 cm)
- a copper strip (2 cm × 15 cm)
- salt solution
- a 250 mL beaker
- 2 pieces of magnet wire (each 20 cm long)
- 2 alligator clips or clothespins

What to Do

1. Make a setup like that shown below.
2. Add enough salt solution to cover about half of the metal strips. Connect the strips to the galvanometer as shown.

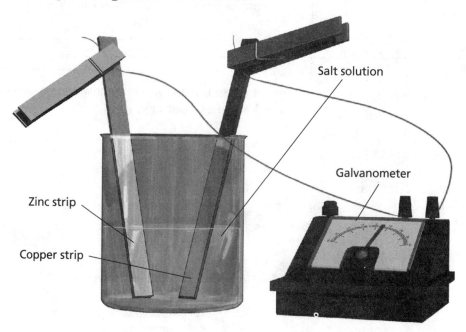

Illustration also on page 311 of your textbook

3. Observe the galvanometer. Record the highest reading reached.

_____

What happens to the reading?

_____

_____

_____

_____

_____

_____

Observe each electrode carefully. What happens to each electrode?

_____

_____

_____

_____

_____

_____

_____

4. Add enough salt solution to fill the beaker.

5. Record the galvanometer reading once again.

_____

_____

Questions

1. How would you connect two wet cells to get more current?

_____

_____

_____

_____

Chapter 14

Draw a sketch showing your answer.

2. What is one way of increasing the current in a wet cell? How would you explain this?

_____

_____

_____

3. What other factors might be altered in a wet cell to increase its current output?

_____

_____

_____

4. What energy changes take place in the operation of dry and wet chemical cells?

_____

_____

_____

EXPLORATION 2

Commercial Electric Cells, page 312

Cooperative Learning Activity	
Group size	2 to 3 students
Group goal	to compare wet and dry cells and to analyze a variety of dry cells
Individual responsibility	Each member of your group should choose a role such as artist, recorder, or director.
Individual accountability	Each group member should be able to write a summary of the activity to explain what he or she learned and what questions he or she still has about cells.

In this Exploration you will examine some common chemical cells. You have probably seen most of them. You may have even wondered how they work. Here's your chance to find out.

Part 1: More Dry Cells

Dry cells were invented to overcome the disadvantages of wet cells. However, dry cells are not really dry. Rather, the solution in them (the electrolyte) is blended with other substances to make it thick and pasty.

1. Look at the cells pictured below. Which would you use in a standard-sized flashlight?

_____

in a penlight?

_____

in a watch?

_____

Several types of
chemical cells
and batteries

Photo also on page 312 of your textbook

Name _____ Date _____ Class _____

2. The voltage of each cell on the previous page is marked. Notice how several cells of different size have the same voltage. How can that be?

_____

_____

_____

What does *voltage* mean to you?

_____

_____

_____

3. Look at the graph below and answer the questions that follow.

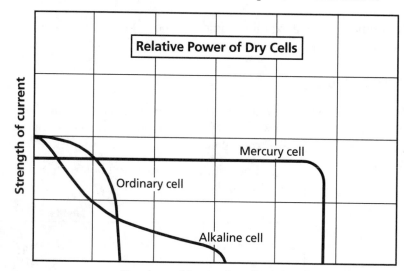

Graph also on page 313 of your textbook

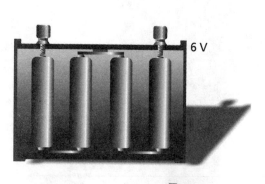

6 V

9 V

D

Which type of cell gradually "winds down"?

_____

Which type loses power quickly?

_____

Which type of cell would you probably use to power devices that require a steady current?

_____

4. Use the descriptions on the next page to help you label the parts of the batteries shown on this page and the next.

Illustration also on page 313 of your textbook

Exploration 2 Worksheet, continued

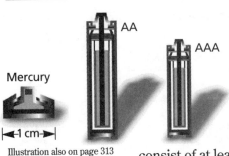

Mercury

|◄–1 cm–►|

Illustration also on page 313
of your textbook

Like all chemical cells, the ordinary dry cell has two electrodes, or conductors, and an electrolyte solution. The *positive electrode* has two parts—a *graphite rod* in the center of the cell and a mixture of *manganese oxide* and *powdered carbon* surrounding the graphite rod. The *negative electrode* is zinc; it makes up the sides and bottom of the cell. The *electrolyte* fills the space between the electrodes. It consists of *ammonium chloride* paste. At the top of the cell is an *insulator*. *Batteries* consist of at least two *individual cells* joined together by *conducting strips*.

In the mercury cell, the *positive electrode* consists of a small block of zinc. The *negative electrode* is a layer of *mercury oxide*. The *electrolyte* is *potassium hydroxide*.

5. The *alkaline cell* differs from an ordinary dry cell in two major ways. First, the negative electrode is made of spongy zinc. Second, the electrolyte is potassium hydroxide, a strong base. What effect do these differences have on the power output of the alkaline cell?

_____

_____

Chapter 14

Part 2: Other Cells

1. A powerful surge of electric current is needed to crank an automobile engine. This surge of current is provided by a group of cells joined together in a battery. Study the drawing below to discover or infer the answers to the questions on the next page.

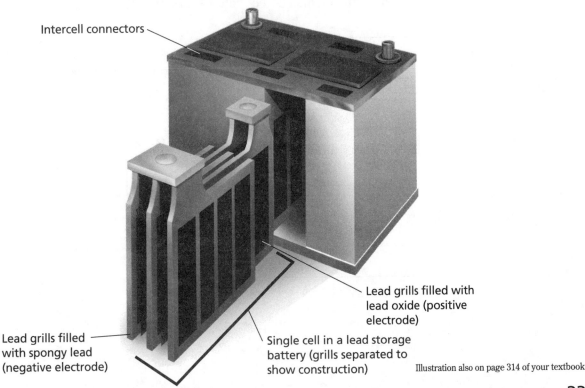

Intercell connectors

Lead grills filled with
lead oxide (positive
electrode)

Lead grills filled
with spongy lead
(negative electrode)

Single cell in a lead storage
battery (grills separated to
show construction)

Illustration also on page 314 of your textbook

Exploration 2 Worksheet, continued

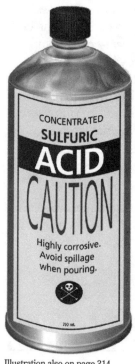

CONCENTRATED
SULFURIC
ACID
CAUTION

Highly corrosive.
Avoid spillage
when pouring.

750 mL

Illustration also on page 314
of your textbook

a. What substances make up (1) the two electrodes and (2) the electrolyte in a car battery?

_____

_____

b. Why is such a large battery needed for a car?

_____

_____

2. Automobile batteries have a limited life span, and not all batteries last the same amount of time. Why do batteries wear out?

_____

_____

_____

Why do some wear out sooner than others?

_____

_____

_____

_____

3. Research "maintenance-free" batteries. How do they work?

_____

_____

_____

How are they different from standard batteries?

_____

_____

_____

_____

4. For many applications, the *nickel-cadmium* cell is replacing both lead-acid batteries and dry cells. Find out how this type of cell works.

_____

_____

_____

_____

_____

_____

5. Unlike dry cells, nickel-cadmium cells and lead-acid batteries can be *recharged.* What does this mean?

_____

_____

_____

_____

How is this property useful?

_____

_____

_____

_____

Chapter 14

EXPLORATION 3

Moving Magnets and Wire Coils, page 315

Cooperative Learning Activity	
Group size	2 to 3 students
Group goal	to demonstrate how a current can be created by the relative motion of magnets and wire coils
Individual responsibility	Each member of your group should choose a role such as coil monitor, magnet mover, or meter reader.
Individual accountability	Each group member should be able to present the group's experimental design for step 6(c) to the class.

Part 1: Building Your Own

You Will Need

- a commercial galvanometer
- a 150 cm length of magnet wire
- a cardboard tube
- a strong bar magnet
- sandpaper

What to Do

1. Sand the enamel off of the last 2 or 3 cm of the ends of the magnet wire. Wrap the magnet wire around the tube to make a coil as illustrated on page 315 of your textbook. Attach the bare ends of the wire to a commercial galvanometer.

2. While watching the galvanometer, move a bar magnet into the coil, hold it there for a moment, and then remove it. Is the galvanometer needle affected?

   _____

   _____

3. Repeat step 2 several times, moving the magnet at different speeds. What do you observe?

   _____

   _____

   Does moving the magnet into the coil have a different effect on the galvanometer than moving the magnet out of the coil? Explain.

   _____

   _____

   What might this suggest about the direction of current flow through the coil?

   _____

   _____

Chapter 1

Exploration 3 Worksheet, continued

4. Disconnect the galvanometer and move the magnet to see whether the magnet itself is affecting the galvanometer. Record your observations.

_____

5. Connect the galvanometer to the coil again. This time, hold the magnet still, but pass the coil over the magnet. What do you observe?

_____

6. Use your observations to answer the following questions:

a. How can a magnet help produce electricity?

_____

_____

b. How is the direction of the current affected by the motion of the magnet?

_____

_____

c. How does the speed of the magnet's motion affect the amount of electricity generated?

_____

_____

d. What changes in forms of energy occur in this investigation?

_____

_____

_____

_____

e. Could a stationary magnet ever produce electricity? Explain.

_____

_____

_____

f. Would current still be generated in the wire if the wire were broken at some point? Why or why not?

_____

_____

_____

Part 2: Francesca's Experiment

Francesca devised an experiment to answer questions raised by Part 1 of this Exploration. She started with a wire coil of 15 turns. Illustrations (a) through (c) show the galvanometer readings she recorded. Then she used a coil with twice as many turns. Illustrations (d) and (e) show these readings.

Francesca tried each part of this experiment three times and obtained similar results each time. What conclusions do you think she drew for each part? The illustrations below provide some hints.

Illustrations also on page 316 of your textbook

a. **b.** Magnet moved inside coil **c.** Magnet moved faster inside coil

d. Magnet moved at the same speed as in (c)

e. Magnet moved out of and away from coil at the same speed as in (d)

Analysis

1. When a magnet is moved inside a coil of wire, _____ is detected in the wire, which _____ its direction when the magnet is moved in the opposite direction inside the coil.

2. A larger current is produced if _____ or if _____.

3. Suppose Francesca moved the magnet into and out of the coil 15 times in a minute. What would happen to the current?

_____

How many times per minute would the current go first in one direction and then in the opposite direction?

_____

_____

_____

_____

4. What do you think would happen if Francesca held the magnet stationary and moved the wire coil instead? Why?

_____

_____

Part 3: A Related Experiment

Francesca made an important discovery: When a magnet is moved through a coil of conducting wire, electricity is generated. Both the number of coils and the speed of movement of the magnet affect the amount of current produced. Francesca also discovered that the direction in which the magnet was moved made a difference. If the magnet was moved in one direction, current flowed one way. If the magnet was moved in the other direction, current flowed the other way. Let's examine the findings of a related experiment.

But before you begin, think a little bit about how a magnet exerts its influence. Does the magnet have to touch something to have an effect, or does its force act through space? Look at the photo at the upper right on page 317 of your textbook. It shows a magnet on which iron filings have been sprinkled. Do you see evidence that (a) the iron filings have been attracted and that (b) the *magnetic force* is exerted through space along curved paths? We call these paths *magnetic lines of force*.

Look at the series of illustrations on page 317 of your textbook, which represent the results of the experiment. The wire is being moved while the magnet is held stationary. The arrows between the north and south poles of the magnet represent the lines of magnetic force.

Analysis

1. How do the results compare to those of Francesca's experiment?

_____

_____

2. What happens when the wire is momentarily motionless as it changes direction, as in (a)?

_____

3. What happens when the wire is moved parallel to the magnetic lines of force, as in (d)?

_____

4. What role do the magnetic lines of force appear to play in the generation of electricity?

_____

_____

Chapter 14

Generators and Motors   Teacher's Notes

This worksheet corresponds to Transparency 47 in the Teaching Transparencies binder.

Suggested Uses	Use as a visual aid with either of the topics listed below.

Generators—Small and Large, page 318

Moving Coils—Revolutionary! page 341

Use the transparency to extend the study of electric generators and motors.

Use the transparency with the transparency worksheet on the next page for reteaching or review. Please note: The worksheet is a reproduction of the actual transparency with certain labels omitted. For answers to that worksheet, see the transparency.

Possible Extension Questions

1. How is this electric generator different from the generators studied earlier in the chapter?

2. What happens to the direction of the current when the coil turns past the vertical direction?

3. Why is the commutator in the motor split into two halves?

4. How could the motor be used to do work?

Answers to Extension Questions

1. It uses a coil instead of a wire loop to generate current.

2. The direction reverses.

3. Splitting the commutator into two halves causes the current to flow in the proper direction through the loop. When the loop turns past the vertical position, each of the brushes comes into contact with a different half of the rotating commutator.

4. The armature could be attached to a drive shaft that would turn and do work.

Generators and Motors

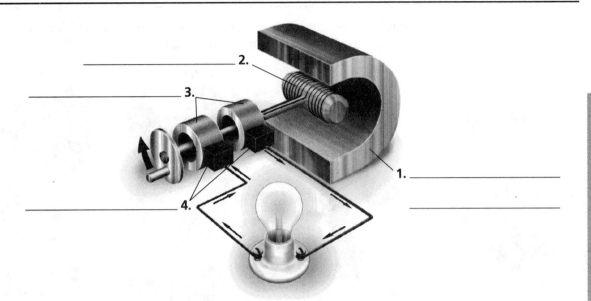

Electric Generator

Chapter 14

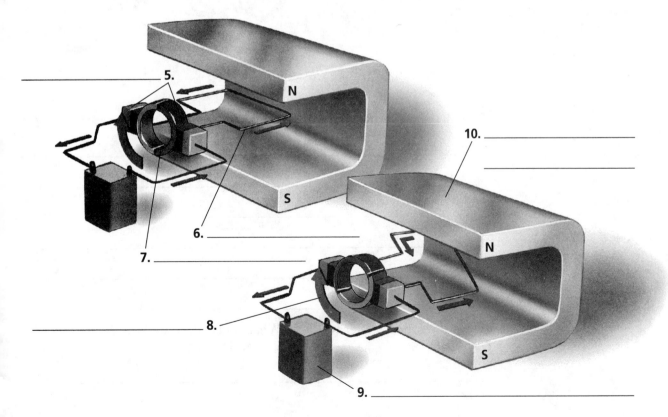

Electric Motor

Challenge Your Thinking, page 324

1. Generator X The sequence of pictures below shows a type of generator in action. Use the pictures to help you answer the questions that follow.

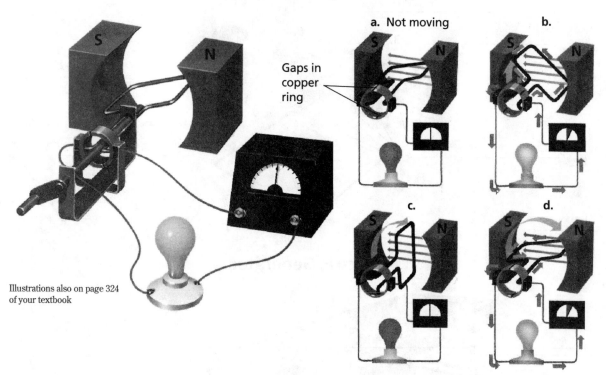

a. Not moving

Gaps in copper ring

Illustrations also on page 324 of your textbook

a. Examine the construction of this generator. How does this generator work?

_____

_____

_____

_____

_____

_____

b. Study the galvanometer readings. What kind of current does this generator produce?

_____

c. How does this generator differ from the generator shown on page 320 of your textbook?

_____

_____

_____

_____

_____

_____

d. Explain what is happening in each illustration in the sequence.

_____

_____

_____

_____

_____

_____

_____

2. Current Events

Electricity is related in some way to each of the following energy forms: light, heat, magnetic, chemical, kinetic, and pressure. Identify the relationship among the energy forms in each of the following converters:

- dry cell   _____

- solar cell   _____

- wet cell   _____

- light bulb   _____

- generator   _____

- piezoelectric
  crystals   _____

- thermocouple   _____

Chapter 14

3. Play It Either Way

A light bulb can use either AC or DC. Hypothesize why this is so.

_____

_____

_____

_____

4. Irregular Exercise

Stan made a jump rope out of a loop of wire and then performed the activity pictured. The galvanometer showed that a current was being generated.

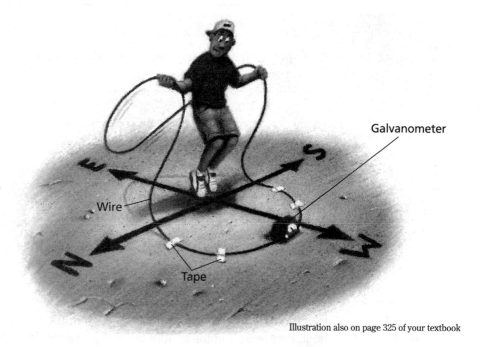

Illustration also on page 325 of your textbook

a. Explain what happened. (Hint: What makes a compass work?)

_____

_____

_____

_____

b. Did Stan generate direct current or alternating current? Explain.

_____

_____

_____

_____

**Correction/
Completion**

1. Complete the following sentence:

   A(n) _____ is a unit of measure equal to 1 cycle per second.

**Short
Responses**

2. What is the difference between a dry cell and a wet cell?

   _____

   _____

   _____

   _____

3. Match each energy converter below with the description of its corresponding energy conversion.

   a. solar cell

   b. generator

   c. piezoelectric crystal

   d. thermocouple

   _____ A combination of mechanical energy and magnetism can produce very large electric currents.

   _____ Materials are stretched or squeezed to produce a current.

   _____ Heating two different metals together produces a current.

   _____ Energy from sunlight is converted into small amounts of electric current.

**Illustration for
Interpretation**

4. Name at least two ways in which this setup could be changed to produce more current.

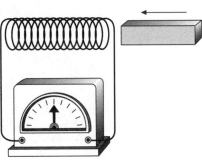

   _____

   _____

   _____

Chapter 14

Short Essay

5. Imagine riding a bicycle after dark. Name one possible problem with using a bicycle dynamo instead of a battery to power the bicycle's light.

_____

_____

_____

_____

_____

_____

_____

Illustration for Interpretation

6. Will this "tree battery" produce continuous electric current? Explain why or why not.

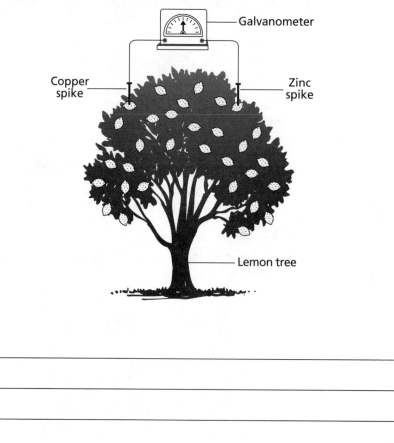

_____

_____

_____

_____

_____

EXPLORATION 1

A Conduction Problem to Investigate, page 329

Cooperative Learning Activity		Safety Alert!
Group size	3 to 4 students	
Group goal	to identify factors that affect the conductivity of a wire	
Individual responsibility	Each group member should choose a role such as chief investigator, checker, recorder, or materials manager.	
Individual accountability	Each group member should be able to complete Applications of Resistance individually.	

The Problem: Do all wires conduct an electric current equally well? Here are some questions to answer as you investigate this problem.
a. Does the kind of metal affect the transmission of current?
b. What is the effect of having different thicknesses of wire?
c. Does the length of wire influence the current?
d. What happens if a wire resists the flow of current?

You Will Need

- magnet wire
- sandpaper
- thin and thick nichrome wire
- a wooden dowel
- D-cells
- a flashlight bulb
- newspaper
- a coin
- a rubber band
- wire cutters

Part 1: Investigating Questions (a) and (b)

What to Do

1. Prepare equal lengths of the three wires that you will test. Sand the enamel off of the last 2 or 3 cm of each end of the magnet wire.

2. Set up the circuit as shown at right using one of the three wires. Observe the brightness of the light.

3. Do the same for each of the other wires. Observe the bulb in each case. Does the intensity of the light vary? _____ Double-check your results.

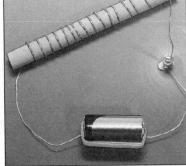

Photo also on page 329 of your textbook

Conclusions

1. Is it easier for a current to flow through thin nichrome wire or thin copper wire?

HRW material copyrighted under notice appearing earlier in this work.

2. Is it easier for a current to flow through thin nichrome wire or thick nichrome wire?

_____

3. How might you explain your observations?

_____

_____

_____

_____

_____

_____

Part 2: Investigating Question (c)
What to Do

1. Vary the length of the thin nichrome wire in the circuit by placing the contact wires at different points along the wire, as shown at left.

2. Observe the bulb.

Conclusions

1. Is it easier for a current to pass through a long piece or a short piece of nichrome wire?

_____

2. How might you explain this observation?

Illustration also on page 329 of
your textbook

_____

_____

_____

_____

Interpreting Parts 1 and 2

Some wires do not allow electric charges to move through them as readily as do other wires. In other words, these wires offer more *resistance* to the flow of the charges. Which offers more resistance:

nichrome or copper wire? _____

thin or thick wire? _____

long or short wire? _____

Resistance can be compared to friction. In what ways do you think they are similar?

_____

_____

_____

Friction

1. Flick a coin across a table with your finger. It moves a little, slows down, and stops. What causes the coin to slow down?

_____

_____

2. Where does the kinetic energy of the moving coin go?

_____

_____

3. Here's how to find out. Place your finger firmly on the coin and rub it back and forth a dozen times or so against a table top. Now touch the coin to your chin. What kind of energy was produced?

_____

4. What caused it to be produced?

_____

Resistance

Resistance is like friction. Electrons flow because they receive electric energy from a cell. As the electrons flow through a piece of nichrome wire, the wire resists the flow of the electrons—in much the same way that the nails resist the rolling marbles in the photo at left.

Photo also on page 330 of your textbook

1. If the nails were a little closer together, how would this affect the rolling marbles?

_____

2. Would this situation represent a wire with more resistance or less?

_____

3. What form of energy do you predict will be produced from the electric (kinetic) energy of the electrons as they slow down?

_____

You will check your prediction in Part 3 of this Exploration.

Chapter 15

Part 3: Investigating Question (d)
What to Do

1. Connect the circuit as shown on page 330 of your textbook.

2. Wrap the wire with newspaper, and then watch it for 1 minute.

3. Now disconnect the circuit, unwrap the newspaper, and carefully touch the wire.

Conclusions

1. When a wire resists the flow of current, what happens?

   _____

2. What energy change is taking place in the nichrome wire?

   _____

Any conductor that offers considerable resistance is called a **resistor** and is represented by the symbol ‑⋁⋁‑. In the space below, use circuit symbols to draw a circuit containing two D-cells, a coil of nichrome wire, a switch, and a bulb.

Applications of Resistance

You have found that resistors produce heat, and you have inferred that they reduce the flow of current. Both of these characteristics are useful.

- Resistors are used to produce heat in appliances like toasters or irons. Appliances that produce heat when an electric current flows through them are called *thermoelectric* devices.

  Why is this a good name for them?

  _____

  _____

  _____

Make a survey of the thermoelectric devices that are used in your home. How much power do they use? (This will be the number, in watts, marked on the appliance.)

_____

_____

Exploration 1 Worksheet, continued

- Resistors also reduce current flow. A *variable resistor*, such as the one used in Part 2 of this Exploration, varies the current. Such resistors are common—the volume control on a radio is one example.

 Here is a project for you to try. Turn an ordinary graphite pencil into a usable rheostat (device for varying the current). What might you use it for?

_____

_____

_____

_____

_____

A graphite-pencil rheostat

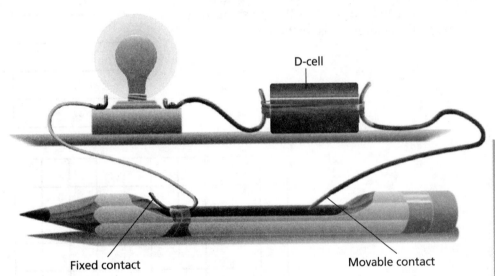

D-cell

Fixed contact

Movable contact

Illustration also on page 330 of your textbook

How does it work?

_____

_____

_____

_____

_____

_____

_____

Chapter 15

Things Are Heating Up

Do this activity after completing Exploration 1 on page 329 of your textbook.

In an experiment to measure the amount of resistance in a circuit, Ian noticed that as the voltage increased, so did the resistance. He also noticed that the temperature of the filament in the bulb increased with the increased voltage. Ian wondered if temperature increase had any effect on the resistance in a circuit. He did some research and found the following data, which show the effect of increased temperature on the resistance of a length of tungsten wire. The filament in the bulb is made of tungsten. Resistance is measured in units called ohms. Plot Ian's data on the grid provided, and then answer the questions on the next page. Don't forget to label the *x*- and *y*-axes.

Temperature (°C)	Resistance (ohms)
27	549
127	783
227	1025
327	1285
427	1562
527	1845
627	2042
727	2421
827	2713
927	3008

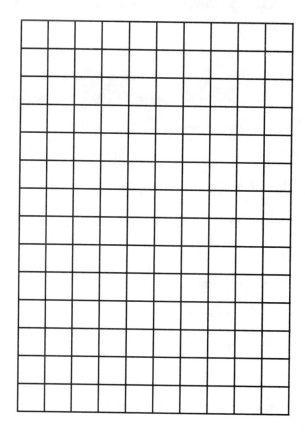

1. What does the graph show? Do you think that there is a relationship between resistance and temperature?

_____

_____

_____

_____

2. Suppose that the data on the previous page is for a piece of tungsten wire 1 m long. What would be the result of doubling the length of the wire? Why do you think this would happen?

_____

_____

_____

3. What do you think would happen to the resistance if the temperature of the wire was cooled toward –273°C? (Note: –273°C is the lowest temperature that matter can have. This temperature is also known as absolute zero.)

_____

_____

You would not want zero resistance in a light bulb filament because the filament would not glow. For what uses would zero resistance be beneficial?

_____

_____

_____

_____

_____

Chapter 15

Chapter 15
Transparency Worksheet

Series and Parallel Circuits   Teacher's Notes

This worksheet corresponds to Transparency 49 in the Teaching Transparencies binder.

Suggested Uses	Use as a visual aid with the topic listed below.

Making Circuits Work for You, page 331

Use the transparency to extend the study of series and parallel circuits.

Use the transparency with the transparency worksheet on the next page for reteaching or review. Please note: The worksheet is a reproduction of the actual transparency with certain labels omitted. For answers to that worksheet, see the transparency.

Possible Extension Questions

1. Trace the path of electric current through the series circuit. Is more than one path possible?

2. What happens to the current when a series circuit is broken at some point?

3. Why does a broken filament in one of the bulbs in the series circuit stop the current?

4. How many different paths are possible for the current in the parallel circuit?

5. What happens to the current in the parallel circuit if one path is broken?

Answers to Extension Questions

1. The path starts at the negative terminal of the battery, goes around the loop, and returns to the battery at the positive terminal. This is the only possible path.

2. The current stops at the point where the circuit is broken.

3. The filament of the bulb is actually a part of the circuit. If the filament is broken, the circuit is incomplete.

4. There are three.

5. The current continues to flow along the other paths.

Name _____ Date _____ Class _____

Series and Parallel Circuits

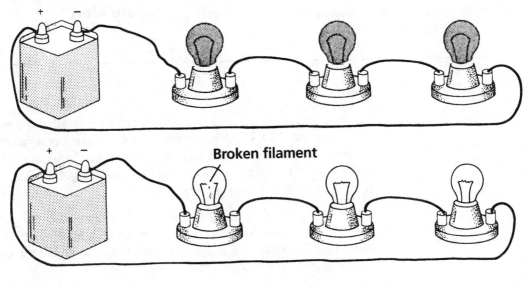

Broken filament

1. _____

2. _____

Chapter 15

EXPLORATION 2

Constructing Circuits, page 331

Your goal	to build and test a number of simple yet functional circuits	Safety Alert

Part 1: Exploring

You Will Need

- 6 pieces of copper wire (each 10 cm long)
- masking tape
- 3 D-cells
- 3 flashlight bulbs in holders
- 2 contact switches

1. You can devise any circuits you wish. Use some or all of the equipment listed at left to construct your circuits. Make any arrangements desired. If you want to use more than one dry cell and do not have a holder, you can use masking tape to hold them together. If the bulbs light up, you have complete circuits.

   Note: Dry cells are quickly drained if left in a closed circuit without something to provide resistance (for example, a bulb). Placing a switch in the circuit and keeping it open until you check the circuit's operation will help to conserve the cell's energy.

2. After constructing your circuits, draw them in your ScienceLog using circuit symbols.

3. Suppose that a certain number of electrons flow out of the cell(s) in a given time. In your ScienceLog, describe the path(s) taken by these electrons.

Part 2: Solving Circuit Problems

The following are a series of circuit problems. What arrangement of circuit components would you make to accomplish the functions described in each problem?

First, make a circuit diagram of your proposed solution in your ScienceLog. Then construct each circuit.

1. A circuit that lights two bulbs, A and B, when the switch is closed. If either bulb burns out or is unscrewed, the other bulb goes out too. This type of circuit is called a **series circuit**.

2. A circuit that lights two bulbs, A and B, when the switch is closed. If either bulb burns out or is unscrewed, the other bulb stays lit. This type of circuit is called a **parallel circuit**.

3. A circuit that contains three bulbs, A, B, and C. If A is unscrewed, then B and C go out. If B is unscrewed, A and C stay lit. If C is unscrewed, A and B stay lit.

Exploration 2 Worksheet, continued

4. A circuit with two bulbs and two switches, P and Q. When P and Q are closed, both bulbs light up. If either switch is opened, neither bulb lights up.

5. A circuit that contains two switches and two bulbs. If both switches are open, neither bulb lights up. If either one of the switches is closed, both bulbs light up.

6. Analyze your findings.

   a. Identify the series and parallel circuits in each of your designs.

   _____

   _____

   _____

   _____

   _____

   _____

   b. Is there any parallel circuitry in the room where you are now? How could you find out without having to expose any wiring?

   _____

   _____

   _____

   _____

Part 3: Current Questions to Investigate

You Will Need

- 6 light bulbs
- 3 contact switches
- 3 D-cells
- copper wire
- wire cutters

Remember: The brightness of the light bulb is a measure of the amount of current flowing.

What to Do

Construct each of the circuits shown in the table on the next two pages, and record the results in the right column. Make certain that switches are included in the circuits you construct.

Putting It Together

On the next page are a number of statements with options. Each statement, with the correct option, is a valid conclusion for one of the four experiments in the table on the next two pages. Choose appropriate conclusions for each experiment and include them in your table.

a. Connecting bulbs one after another in a circuit (decreases, increases) the amount of current flowing in the circuit.

b. If the number of cells is increased in the circuit, the amount of current is (decreased, increased).

c. If more bulbs are connected in series in a circuit, the resistance of a circuit is (decreased, increased).

d. If two bulbs are placed in a branched circuit rather than in an unbranched circuit, the current through the battery is (decreased, increased).

e. The current flowing through each bulb in a parallel circuit is (greater than, less than) the current flowing through each bulb in a series circuit.

f. The resistance of a circuit is decreased when the bulbs are placed in (series, parallel).

g. The resistance of a circuit is (more, less) with three bulbs in series than with two bulbs in parallel connected to a third in series.

Question	Experimental design	Results/conclusions of experiment (Select from list above)
1. How is the amount of current affected by the number of cells in a circuit?	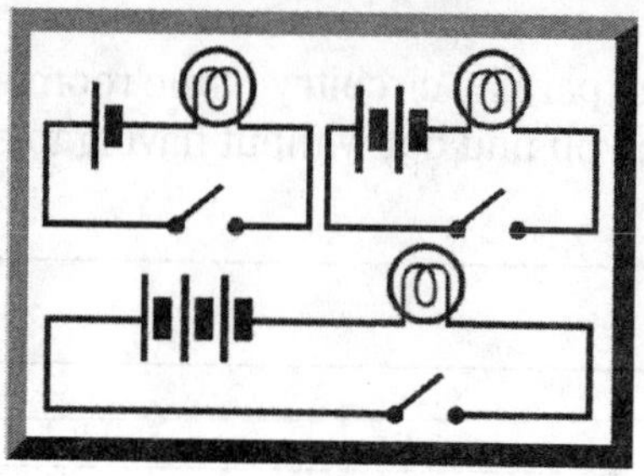	**Result:** **Conclusion(s):**
2. How is the amount of current affected by the number of bulbs connected *in series*, that is, one right after the other?	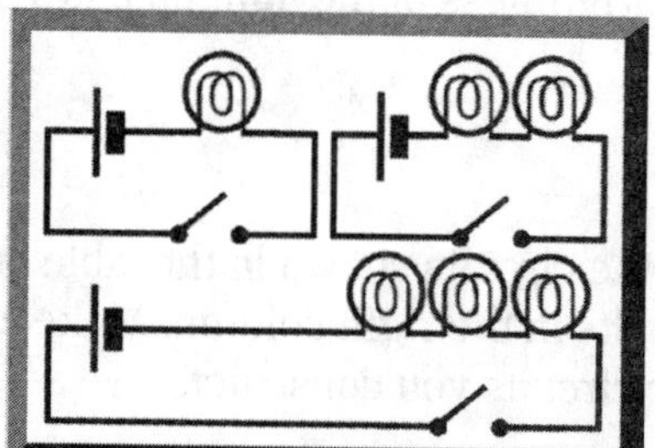	**Result:** **Conclusion(s):**

Name _____ Date _____ Class _____

Question	Experimental design	Results/conclusions of experiment (Select from list on the previous page)
3. How does the current flowing through each bulb in a *parallel,* or branched, circuit compare with the current flowing through each bulb in a series circuit?	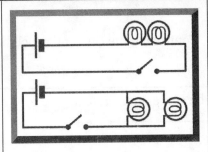	**Result:** **Conclusion(s):**
4. What difference is there between the current flowing through the battery when two bulbs are connected in series and the current flowing through the battery when two bulbs are connected in parallel?	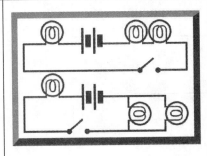	**Result:** **Conclusion(s):**

Chapter 15

Circuit Mastery

Try this activity after you complete Lesson 3, Controlling the Current, which begins on page 335 of your textbook.

If you can complete these drawings, you will be current on circuits.

a. Modify the circuit by adding what is needed to light the bulb.

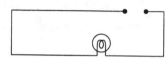

b. Modify the circuit by adding what is necessary to keep the cell from being quickly drained of its electrical energy.

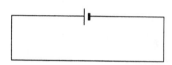

c. Modify the circuit by adding what is necessary to turn each light on and off independently.

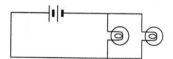

d. Modify the circuit by adding what is necessary to turn the light on with switches located in two different places.

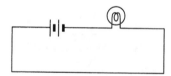

e. Modify the circuit by adding something to the circuit to control the brightness of the light.

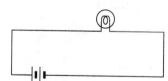

f. Modify the circuit by adding a switch that will turn all of the bulbs on and off. Then add switches so that bulbs *B* and *C* can be controlled individually while bulb *A* is on.

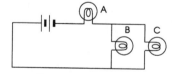

g. Draw two switches to operate the circuit so that if the bulb is lit when one switch is closed, the other switch will be able to open the circuit—and vice versa. Add any wiring that may be needed.

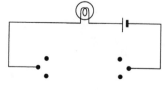

EXPLORATION 4

Constructing an Electromagnet, page 338

Your goal	to construct an electromagnet strong enough to hold up a number of washers	**Safety Alert!**

You Will Need

- a paper clip
- washers
- some light, insulated wire
- 2 D-cells
- an iron spike
- a switch

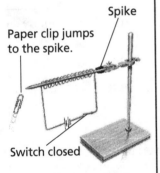

Paper clip jumps to the spike.

Spike

Switch closed

Illustration also on page 338 of your textbook

What to Do

Get together with one or two classmates. Your task will be to make a functioning electromagnet and then to determine how its strength can be increased.

Making the Electromagnet

1. Use the materials illustrated to make an electromagnet capable of supporting a paper clip from which several washers are hanging.

2. Sketch your electromagnet. Then label and trace the path of the current.

3. Are the spike, paper clip, and washers part of the circuit? _____

4. How many washers can your first design hold? _____

Increasing the Electromagnet's Strength

How can you make your electromagnet hold more washers? Take some time to discuss the following:

- factors or variables that might be altered
- ways to measure the magnet's strength
- safety precautions
- the apparatus

    Get your design approved by your teacher. Then assemble the necessary apparatus, do the experiment, record the results, and draw conclusions based on your results. Record your results in your ScienceLog and share your results with others.

Chapter 15

Name ______________________ Date ____________ Class ____________

Using Electrical Units

Chapter 15
Activity Worksheet

Do this activity with Lesson 5, How Much Electricity? on page 343 of your textbook.

Complete the table below.

What we are measuring	Unit	Definition	Problems
quantity of charge	1 coulomb	6.24 quintillion electrons	How many electrons are in 5 C of charge? 31.2 quintillion
	1 ampere		**a.** For 5 A, how many coulombs of charge pass in 2 s? ______ What is the current? ______ **b.** If 20 C of charge pass in 2 s, what is the current? ______ **c.** If 4 A of current flow for 5 s, what is the quantity of charge? ______
	1 volt		**a.** How much energy per coulomb is given by a 6 V battery? ______ **b.** How much energy is given to 5 C of charge by a 6 V battery? ______ **c.** If 6 J of energy are given to 3 C of charge, what is the strength of the battery? ______
	1 watt		**a.** How many joules of energy are used in 1 s by a 600 W iron? ______ **b.** How many joules of energy does a 2000 W heater use in 1 min.? ______ **c.** If a radio uses 1000 J in 10 s, what is its power? ______
	1 kilowatt		**a.** 200 W equals how many kilowatts? ______ **b.** 500 kW equals how many watts? ______

Name _____ Date _____ Class _____

A Classical Current Demonstration, page 300

Here is a demonstration similar to one first done hundreds of years ago. Make a setup like the one shown below, and try it yourself. Follow the steps closely.

You Will Need

- a plastic strip, such as a plastic ruler
- plastic wrap
- scissors
- 30 cm of thread
- a wheat puff
- a wire clothes hanger
- a nail
- a paper cup
- a vinyl strip
- flannel cloth

Coat hanger

Thread

Notched paper cup

Wheat puff

Charged plastic strip

Illustration also on page 300 of your textbook

Nail

1. Vigorously rub a plastic strip, such as a plastic ruler, with plastic wrap.

2. Quickly touch the strip or ruler to the end of the nail. What happens to the wheat puff?

_____

_____

3. Rub the plastic strip again and bring it close to the end of the nail without quite touching it. Observe the wheat puff.

_____

4. Repeat the experiment using a vinyl strip rubbed vigorously with flannel cloth. What happens to the wheat puff now?

_____

How do you explain the events in this demonstration?

_____

_____

_____

_____

_____

Name ______________________ Date __________ Class __________

Chapter 13
Resource Worksheet

A Theory of Charged Particles, page 301

Scientists explain events such as those seen in A Classical Current Demonstration on page 300 of your textbook in this way:

1. When one material is rubbed against another, friction causes charged particles to move from one material to the other. The accumulated charge is indicated by the charged material's ability to attract lightweight or finely powdered substances. When did this happen in the previous demonstration?

2. There are two kinds of charged particles: positively charged particles and negatively charged particles. Why might you conclude that the charges on the plastic strip are different from those on the vinyl strip?

3. Objects that have the same kind of charge tend to repel one another. Objects that have different charges tend to attract one another. When did you observe these effects in the demonstration?

4. Charged particles pass easily through certain materials, called **conductors**, but pass with difficulty or not at all through other materials, called **insulators** (nonconductors). What evidence is there that iron is a conductor of electricity?

Using the Theme of Systems

This worksheet is an extension of the theme strategy outlined on page 348 of the Annotated Teacher's Edition. It is also designed as an extension of Chapter 15, Currents and Circuits, which begins on page 326 of the Pupil's Edition.

Focus question	How is current through a circuit similar to blood flow through the human body?

The human circulatory system is responsible for transporting blood throughout the body. Do some research to learn about the circulatory system, and then answer the following questions:

1. The heart acts as a pump, pushing blood through the arteries at high pressure. What part of a circuit acts as an electron pump?

   _____

2. Is the flow of blood through the body a "direct current" or an "alternating current"? Explain your reasoning.

   _____

   _____

3. Would you define the circulatory system as a parallel circuit or a series circuit? Explain your reasoning.

   _____

   _____

4. Current meets with resistance as it flows through a circuit. Where might blood flow experience resistance in the circulatory system?

   _____

   _____

   _____

5. Valves in the heart and veins are similar to diodes, electrical devices that allow current flow in only one direction. Explain this similarity.

   _____

   _____

   _____

   _____

Chapter 15

Name ________________________________ Date ____________ Class ____________

Chapter 15
Activity Worksheet

An Electrifying Magic Square

Try this activity after completing Chapter 15, which begins on page 326 of your textbook.

Select the number of the term in list II that best matches each statement in list I. Place the number in the appropriate space in the magic square. Add the horizontal rows, the vertical columns, and the diagonals to discover the magic number. Not all of the items in list II will be used.

List I

A. A flow of electrical charges
B. One coulomb of charge flowing past a point in a circuit in one second
C. A measure of the force that pushes current through a conductor
D. 1 J/s
E. Actual charged particles flowing in most circuits
F. Unit for electrical charge
G. One joule of electrical energy per coulomb
H. Electrical energy (large quantity)
I. Electrical energy (small quantity)

List II

1. proton
2. ampere
3. joule
4. watt
5. volt
6. electrons
7. voltage
8. coulomb
9. current
10. kilowatt-hour
11. neutrons

A	B	C
D	E	F
G	H	I

Magic number = ____________

Name ______________________ Date ____________ Class ____________

Chapter 15
Review Worksheet

Challenge Your Thinking, page 348

1. Safe Circuit

Many apartment buildings have security doors that can be opened from each apartment. The diagram on page 348 of your textbook shows the circuitry of one such security door. Explain how it works.

____

____

____

____

____

____

2. Go With the Flow

The drawing on the next page from a sixth-grade science book illustrates a lesson about electricity. Write a paragraph to go with this lesson and then fill in the blanks in the diagram on the next page.

____

____

____

____

____

____

____

____

____

____

____

____

____

____

What words should be used to fill in the blanks?

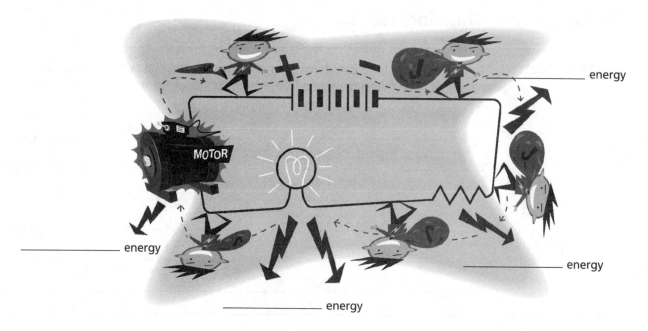

Illustration also on page 348 of your textbook

---

3. The Inside Story
A common type of switch is the *dimmer switch*. Turn the knob clockwise, and the light becomes brighter. Turn the knob counterclockwise, and the light becomes dimmer. Draw a sketch showing the circuit that may be inside.

4. You Be the Teacher

Ms. Alvarado asked her class to design a circuit containing two switches (*S1* and *S2*), two light bulbs (*L1* and *L2*), and one dry cell. She asked them to arrange the circuit so that it does the following:

a. If only *S1* is closed, *L1* will light.

b. If only *S2* is closed, nothing will happen.

c. If *S1* and *S2* are closed, both lamps will light.

Their work is shown below.

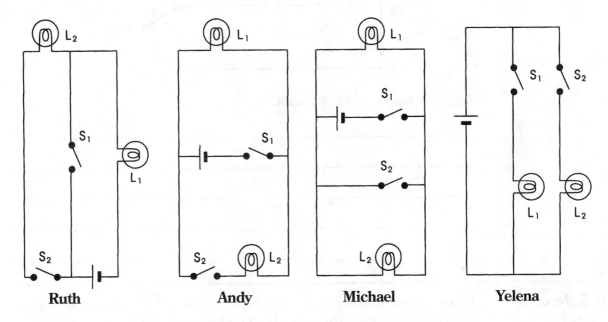

Ruth   Andy   Michael   Yelena

Play the role of Ms. Alvarado, and grade the students' work. Indicate how you know whether each design is right or wrong.

_____

_____

_____

_____

_____

_____

_____

_____

_____

_____

_____

5. The Ol' Double Switcheroo

In the circuit shown below, the upper switch moves from contact A to contact C and the lower switch moves from contact B to contact D.

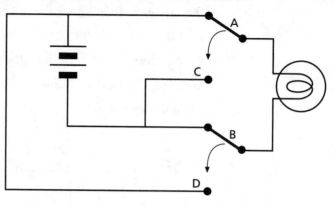

Suggest a function for this circuit.

_____

_____

_____

_____

_____

6. Safe at Home

Fuses are designed to protect household circuits against damage due to electrical overload. Fuses are made of metal alloys that have low melting points. When a circuit carries more current than is safe, the fuse strip melts. How do you think fuses work? (Hint: Think about the effects of resistance.)

_____

_____

_____

_____

_____

_____

_____

_____

_____

_____

Chapter 15
Assessment

Word Usage

1. Use all of the following terms in one or two sentences to show how they are related: *amperes, current, joules, coulombs, energy,* and *volts.*

_____

_____

_____

_____

**Correction/
Completion**

2. The following sentences are incorrect or incomplete. Your challenge is to make them correct and complete.

   a. In a parallel circuit, if one appliance burns out, all of the other appliances stop working.

   _____

   _____

   b. Appliances such as toasters and irons convert _____

   energy into _____ energy because their circuits

   offer a lot of _____.

**Answering by
Illustration**

3. Using the correct symbols, draw a circuit diagram to match this illustration.

Chapter 15

Name ______________________ Date __________ Class __________

Numerical Problem

4. A 400 W hair dryer is connected to a 110 V household current.

 a. How many joules of energy are applied to each coulomb of charge?

 ____

 ____

 b. If the hair dryer is used for 3 minutes, how many joules of energy are used?

 ____

 ____

 c. How many coulombs were required during those 3 minutes?

 ____

 d. If the charge was constant during the 3 minutes, what was the rate of electrical flow per second?

 ____

CHALLENGE 1

Answering by Illustration

5. Draw an illustration to match this circuit diagram.

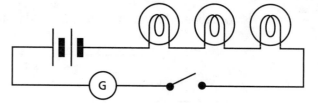

CHALLENGE 2

Short Essay

6. Explain why modern homes use parallel circuits rather than series circuits.

____

____

____

____

____

Electrifying News

Lana Lectron used to be an electrical engineer before she became a reporter, which could explain why she sometimes "electrified" some key words and phrases in the story that follows. See if you can spot the electrical terms, and then use them to fill in the blanks in the glossary that follows.

Dangerous Nic Rome was captured by Detectives Ronald and Rhoda Franklin. They work for the A.C.D.C. (Apprehend Criminals Detective Company). Nic Rome has been charged with the conduction of an attractive 18-year-old coil, Millie Volt, and with the theft of valuable joules.

This is watt appears to have happened. The criminal avoided a battery of policemen by escaping on his kinetic cycle. Spotting him from a parallel road, the detectives proton their hats and jumped into their motor car. In an instant, the thermocouple were in hot pursuit. They forced Nic Rome off the road, causing him to crash into a coulomb of poles. Nic grabbed his weapon and swung it with vicious energy to ampere his arrest. The detectives warded off this series resistance and subdued the potential killer.

Nic Rome is now in solartary confinement in the local chemical cell. His trial will be held in the current court session presided over by the circuit judge. The community is relieved at the capture of the crime generator. As a result, they have agreed to elect Ron and Rhoda Detectives of the Year.

Glossary of Electrical and Magnetic Words

a. Units

1. Unit of electric charge _____

2. Unit of current; one coulomb per second _____

3. Unit of energy _____

4. One joule per coulomb _____

5. One joule of energy used in one second _____

b. Energy and Energy Converters

6. Type of energy possessed by a raised object _____

7. Energy of moving objects _____

8. Device that converts electrical energy into mechanical energy _____

9. Device in which electrical energy is produced by chemical action _____

10. Device in which a current is produced by a magnet revolving near a coil of wire _____

11. Group of cells _____

12. Sensitive device for measuring temperature _____

Electrifying News, continued

c. Circuits

13. The easy passage of electric charges through a
material

_____

14. Path along which electrons flow

_____

15. Type of friction met by electric charges moving
in a conductor

_____

16. Type of circuit in which appliances are connected
one after the other

_____

17. Type of circuit with branches

_____

18. Part of an electromagnet

_____

19. A current that moves in one direction and then
in the opposite direction

_____

20. A current that moves only in one direction

_____

21. Rate of flow of electric charges in a conductor

_____

22. A type of wire with high resistance to the flow
of current

_____

23. One reversal of current

_____

d. Charges and Cells

24. Smallest particle with negative charge

_____

25. Small particle with positive charge

_____

26. Conducting terminals in a cell

_____

27. Given to electrons by a chemical cell

_____

28. Type of cell used to obtain electrical energy from
light energy

_____

e. Magnets

29. Type of force between the north and south
poles of a magnet

_____

30. Parts of a magnet where the strength is greatest

_____

Making Connections, page 350

The Big Ideas   In your ScienceLog, write a summary of this unit, using the following questions as a guide:

1. How may electricity be produced? (Ch. 13)

2. What is the difference between a cell and a battery? (Ch. 13)

3. What parts are common to all chemical cells, and how do they operate? (Ch. 13)

4. How do AC and DC currents differ? (Ch. 14)

5. What factors affect the size of currents in circuits? (Ch. 15)

6. What are some differences between series and parallel circuits? (Ch. 15)

7. What are electromagnets? How can their strength be increased? (Ch. 15)

8. How is the flow of water like that of electricity? (Ch. 15)

9. By what units is electricity measured? How are these units related? (Ch. 15)

Checking Your Understanding

1. Recall your introduction to fuses in Challenge Your Thinking, Chapter 15, on page 349 of your textbook. Suppose a fuse is rated at 15 amps (it melts when the circuit carries more than 15 amps). How many 100 W light bulbs would it take to blow the fuse? Assume a 110 V current. (Don't try this yourself!) You will find questions 6 and 7 on page 346 of your textbook to be helpful.

_____

_____

_____

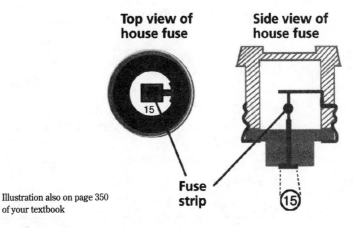

Top view of house fuse

Side view of house fuse

Illustration also on page 350 of your textbook

Fuse strip

2. Below is a diagram of a circuit breaker, a device that mechanically performs the same task as a fuse, breaking the circuit when too much current flows through it. How does this device work?

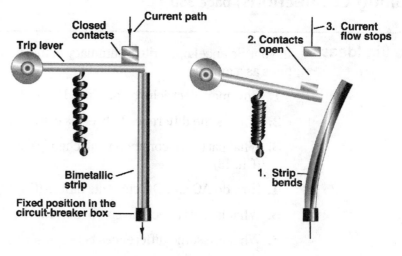

A. Normal current **B. Overload current**

Illustration also on page 351 of your textbook

_____

_____

_____

_____

_____

_____

_____

3. Electricity is often compared to flowing water. Read the following examples and decide whether each suggests high or low amperage, high or low voltage, or any combination of these.

a. the Mississippi River

_____

b. Niagara Falls

_____

c. the blast from the spray-gun nozzle at a do-it-yourself carwash

_____

d. a dripping faucet

_____

Unit 5 Review Worksheet, continued

4. The diagrams below show two different types of microphones.

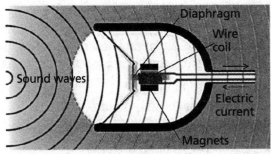
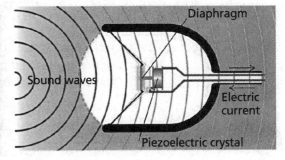

Illustration also on page
351 of your textbook

Use the diagrams and your knowledge of the principles of electricity to explain how each of these microphones works.

_____

_____

_____

_____

_____

_____

_____

5. 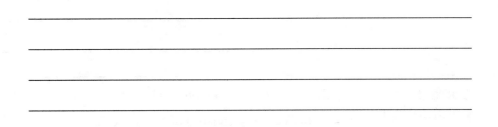 Make a concept map using the following terms: volts, current, electricity, amperes, voltage, coulombs, and charge.

Name ________________________ Date ____________ Class ____________

Unit 5
End-of-Unit Assessment

Word Usage

1. Write a sentence about each of the situations described below. In each sentence, use at least one of these terms: *resistance, dry cells, wet cells, series, kilowatt-hours,* and *circuit.*

   a. The heating element in the toaster glows red.

   ____

   ____

   ____

   b. The meter reader checked the meter on our house today.

   ____

   ____

   ____

Correction/ Completion

2. Correct the statements below.

   a. A battery consists of a single cell.

   ____

   ____

   b. Energy flow is always counterclockwise in an alternating current generator.

   ____

   ____

   c. A series circuit has at least two branches, with the current divided between each.

   ____

   ____

Short Responses

3. Of the following items, circle the ones that would make good conductors of electricity.

plastic ruler	safety pin	plastic comb
wooden pencil	scissors	nail
plastic wrap		

_____

Unit 5 Assessment, continued

_____

4. Match each of the following terms with the proper definition or description.

    a. ampere   _____ a unit of one cycle per second

    b. power   _____ a unit of measure of electrical charge

    c. hertz   _____ 1 C passing a point in 1 s

    d. coulomb   _____ the energy of the charge flowing in a circuit

    e. voltage   _____ the rate at which electrical energy is used

    f. current   _____ flow of charge in a conductor

_____

Illustration for Correction or Completion

5. Something is missing from each of the drawings. Sketch and label the missing parts.

    a. electromagnet

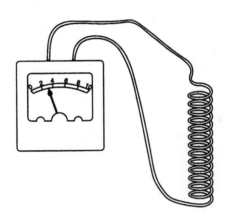

    b. complete circuit

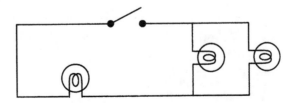

Unit 5

Name _____ Date _____ Class _____

Illustration for Interpretation

6. Use the illustrations below to answer the questions that follow.

Circuit 1

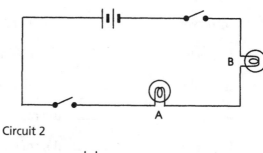

Circuit 2

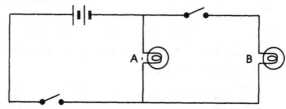

a. How is Circuit 1 different from Circuit 2?

_____

b. If you want to be able to use light A without light B, which circuit would you use?

_____

c. In Circuit 1, does either bulb glow more brightly than the other? If so, which one?

_____

d. In Circuit 2, does either bulb glow more brightly than the other? If so, which one?

_____

Numerical Problem

7. A toaster, electric kettle, and coffee maker are all being operated from the same electrical outlet.

Information: kettle   1500 W
                toaster   800 W
                coffee maker   450 W
                The house operates on a 110 V circuit.

Would a 20 A fuse blow in this case? Show your work.

_____

_____

Graph for Interpretation

8. Use the graph below to answer the questions that follow.

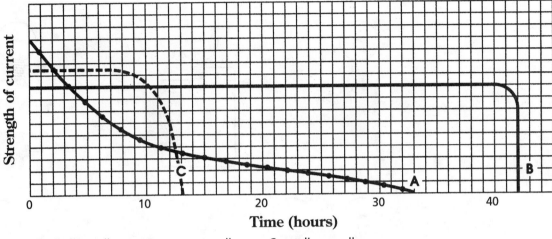

A = alkaline cell   B = mercury cell   C = ordinary cell

a. Which cell has the shortest life?

_____

b. How many hours will the mercury cell last?

_____

c. Which is the strongest cell when new?

_____

CHALLENGE

Answering by Illustration

9. Using the principles of an electromagnet, design and draw a model crane that you could use to pick up thumbtacks from the floor. Clearly label your illustration to show how it works.

Equipment: cells (2) or battery   crank
               wires   nail
               pulley and cable   switch

Illustration for Correction or Completion

10. If necessary, correct each of the following circuits by adding, removing, or changing a wire to make a bulb light. If the circuit is already complete, write a *C* inside it.

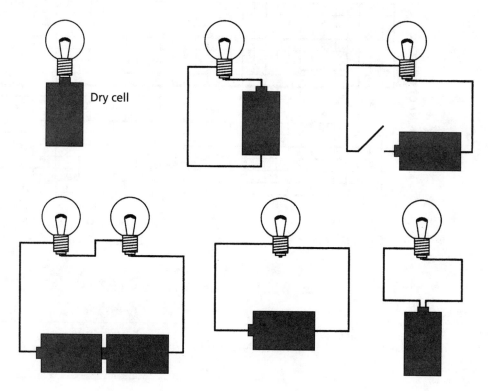

Dry cell

Name ______________________ Date ____________ Class ____________

Unit 5
Activity Assessment

Current Action Teacher's Notes

Overview	Students construct and modify a circuit in order to observe how a change in the direction of electron flow results in a change in the magnetic field.

Materials
(per activity station)

- a strip of aluminum foil (25 cm × 1 cm)
- a D-cell with holder
- 2 lengths of insulated wire (50 cm each)
- a small paper plate
- 2 paper clips
- a magnet

Preparation

Prior to the assessment, equip student activity stations with the materials needed. Remind students that while electricity can be dangerous, the amounts of electricity used in this activity are quite small and unlikely to cause harm.

Time Required

Each student should have 25 minutes at the activity station and 15 minutes to complete the Data Chart.

Performance

At the end of the assessment, students should turn in the following:

- a completed Data Chart

Evaluation

The following is a recommended breakdown for evaluation of this Activity Assessment:

- 40% use of materials to set up a circuit
- 60% ability to make observations and correctly interpret results

Current Action

In this activity you'll investigate magnetic fields in two different types of circuits.

Before You Begin . . .	As you work through the tasks, keep in mind that your teacher will be observing the following: • how you use the materials to set up a circuit • how well you make observations and correctly interpret results Go with the flow!
Task 1	Begin by touching the magnet to the foil. What happens? Record your observation and an explanation in the Data Chart on the next page.

Task 2   Place the magnet flat on the plate. Position the foil over the magnet and secure it to the sides of the plate with paper clips, as shown in the diagram. Connect the wires to the D-cell, and then touch the ends of the wire to the paper clips.

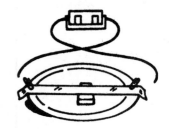

D-cell

Paper clip   Magnet   Aluminum foil

Caution: Do not hold the connection for more than a couple of seconds because the wire will become warm, and the D-cell will drain.

What happens? Record your observations and an explanation in the Data Chart on the next page.

Task 3   Change the direction of electron flow through the foil by reversing the wires to each paper clip. What happens now? Note your observations and an explanation in your Data Chart.

Task 1

Data Chart

Observation	Explanation

Task 2

Observation	Explanation

Task 3

Observation	Explanation

Name _____ Date _____ Class _____

Self-Evaluation of Achievement

The statements below include some of the things that may be learned when studying this unit. If I have put a check mark beside a statement, that means I can do what it says.

_____ I can demonstrate the charging of a material by friction and show that the material has an electrical charge. (Ch. 13)

_____ I can show that there is more than one type of electrical charge. (Ch. 13)

_____ I can explain the following in terms of the existence of negatively and positively charged particles: a negatively charged material, a positively charged material, an uncharged material, a current, what causes a current to flow, and what causes a current to be continuous. (Ch. 13)

_____ I can demonstrate that a current flowing in a wire has magnetic properties. (Ch. 14)

_____ I can construct a simple chemical cell and detect the current flowing from it. (Ch. 14)

_____ I can demonstrate the production of an alternating electric current by a coil and a wire and name the factors that affect the amount of current. (Ch. 14)

_____ Given a clear illustration of the main components of an electric generator or an electric motor, I can explain its operation. (Ch. 14)

_____ I can construct electric circuits containing series or parallel arrangements of cells, lamps, resistors, and switches, and I can use circuit symbols to draw diagrams of these circuits. (Ch. 15)

_____ I can construct an electromagnet. (Ch. 15)

I have also learned to _____

_____

_____

_____

_____

I would like to know more about _____

_____

_____

_____

_____

Signature: _____

Name _____ Date _____ Class _____

Making a Compass

Complete this activity after reading pages S91–S96 of the SourceBook.

You Will Need

- a large sewing needle
- a piece of cork
- a magnet
- tape
- a small dish
- a compass
- water

What to Do

1. Hold the needle carefully by the blunt end. Hold the magnet with your other hand and move one pole along the length of the needle in the same direction at least 25 times. Explain the effect this action has on the needle.

   _____

2. Construct a compass by taping the needle horizontally to the piece of cork. Then float the cork in a small dish of water. Try pointing the needle in different directions and releasing it. What happens? Explain.

   _____

   _____

   _____

3. Use the commercial compass to check the results obtained with your homemade compass. Were your results accurate?

   _____

   _____

4. Name two factors that might cause the results to be different.

   _____

   _____

   _____

   _____

5. Name the three magnets involved in this activity, and describe their alignment.

   _____

   _____

   _____

   _____

   _____

Unit CheckUp, page S103

**Concept
Mapping**

The concept map shown here illustrates major ideas in this unit. Complete the map by supplying the missing terms. Then extend your map by answering the additional question below.

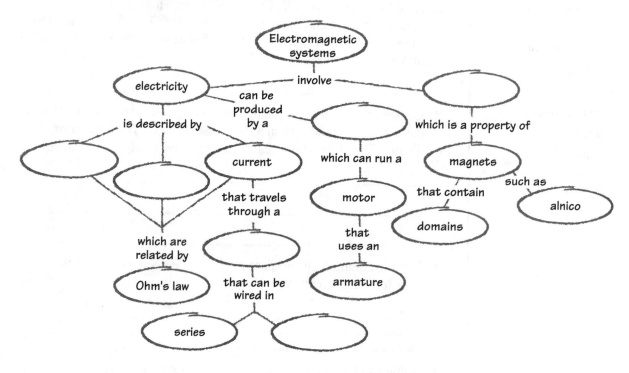

Where and how would you connect the terms *electrons*, *poles*, and *transformer*?

**Checking Your
Understanding**

Select the choice that most completely and correctly answers each of the following questions.

1. An electric current results from

   a. the flow of positively charged particles through a conductor.

   b. the flow of electrons between areas of greater and lesser charge.

   c. the movement of ions through wires.

   d. the flipping of a switch.

2. Which of the following correctly describes the behavior of charged particles?

a. Negatively charged particles attract positively charged particles.

b. Negatively charged particles attract other negatively charged particles.

c. Positively charged particles attract other positively charged particles.

d. Negatively charged particles attract both positively charged and negatively charged particles.

3. Voltage can be compared to

a. flow rate.

b. current velocity.

c. pressure.

d. friction.

4. "The greater the distance between two objects, the weaker their interaction," is a statement that applies to

a. only an electric force.

b. only a magnetic force.

c. both an electric and a magnetic force.

d. neither an electric nor a magnetic force.

5. A generator converts

a. electrical potential into electrons.

b. electrical energy into mechanical energy.

c. direct current into alternating current.

d. mechanical energy into electrical energy.

Critical Thinking

Carefully consider the following questions, and write a response that indicates your understanding of science.

1. Suppose that you have an iron nail and a compass. How could you demonstrate whether the nail was magnetized or not?

_____

_____

_____

_____

_____

_____

2. A certain transformer has a primary coil with 30 turns of wire and a secondary coil with 10 turns; the voltage in the secondary coil is 30 V. What is the voltage in the primary coil? Show your work.

_____

_____

_____

3. Suppose you coupled a generator to an electric motor so that the motor drove the generator and the generator supplied the motor with electricity. Would this system run itself indefinitely? Why or why not?

_____

_____

_____

4. Why is AC, but not DC, easily transformed?

_____

_____

_____

What would happen if you passed DC through a transformer? Explain.

_____

_____

_____

5. All magnets have a *Curie point*, the temperature at which the materials cease to be magnetic. Above that temperature, no magnetism is present. As you know, there is a relationship between temperature and molecular behavior. How does this relationship help to explain the Curie point?

_____

_____

_____

_____

_____

_____

_____

Name ______________________ Date __________ Class __________

Interpreting Photos

Some of the bulbs in this string of lights are not working. What type of circuit—series or parallel—would account for the on-and-off pattern of bulbs shown here? Explain.

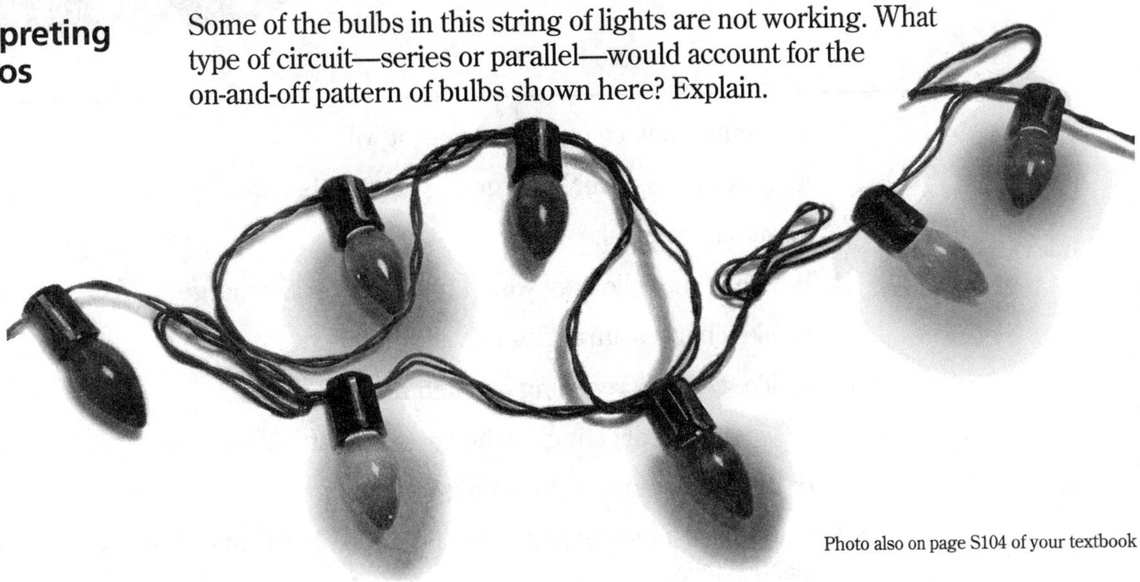

Photo also on page S104 of your textbook

Portfolio Idea

Imagine that you are a scientist in prehistoric times, and you have just discovered magnetism. You are describing its properties to a colleague in a letter. Since you have discovered it well before the modern age of science, most of the words associated with magnetism do not yet exist. In your ScienceLog, describe your discoveries about magnets and magnetism to your colleague without using the terms *magnet*, *magnetism*, *pole*, *magnetic field*, or *lines of force*.

1. If a neutral object loses electrons, it will

   a. become positively charged.   **b.** become negatively charged.

   c. remain neutral.

2. Which of the following would result in a static charge?

   a. like charges attracting one another

   b. like charges repelling one another

   c. electrons collecting on the surface of an object

   d. electrons moving through a conductor

3. Only the distance between two charged objects affects the strength of an electric force.

   a. true   **b.** false

4. According to Coulomb's law, if the distance between two charged objects is decreased, the electric force will

   a. remain the same.   **b.** decrease.   **c.** increase.

5. Voltage is a measure of

   a. current.   **b.** potential difference.   **c.** wattage.   **d.** magnetism.

6. The filament in a light bulb gives off heat in addition to light because of the _____ in the filament.

   a. potential difference   **b.** pressure   **c.** voltage   **d.** resistance

7. Ohm's law states that the _____ increases with increasing _____ but decreases with increasing resistance.

   a. voltage, power   **b.** static electricity, amperage

   c. electric current, voltage   **d.** potential difference, electric current

8. Magnetism is not evenly distributed in a magnet.

   a. true   **b.** false

9. When you cut a bar magnet in half, one half acts as the south pole of the magnet and one half acts as the north pole of the magnet.

   a. true   **b.** false

10. Magnetic lines of force

    a. often cross each other.

    b. are straight lines at the end of a magnet.

    c. have no effect on a compass needle.

    d. can be used to predict which way a compass needle will point.

11. In an atom, magnetic force is caused by

    a. the interaction of protons and electrons.

    b. the motion of unpaired electrons.

    c. the number of paired electrons in the atoms.

    d. electrons orbiting the nucleus.

12. Magnetic domains are _____ in an unmagnetized iron bar.

    a. absent   **b.** randomly arranged

    c. aligned   **d.** None of the above

13. Which material is naturally magnetic?

    a. alnico   **b.** magnequench   **c.** magnetite   **d.** iron

14. Electromagnets are _____ magnets.

    a. temporary   **b.** permanent

15. When using a compass, you are working with two magnets. What two magnets are they?

_____

16. Which is *not* involved in the production of alternating current?

    a. brush   **b.** commutator   **c.** armature   **d.** magnetic field

17. A transformer with two more turns in the secondary coil than in the primary coil

    a. is a step-up transformer.   **b.** will decrease voltage.

    c. cannot produce current.   **d.** Both **a** and **c**

18. A motor can be thought of as the reverse of a generator.

    a. true   **b.** false

19. Match each property on the left with the correct type of force on the right. You may need to use an answer more than once.

_____ works at a distance   **a.** electric

_____ becomes stronger as distance between involved objects increases   **b.** magnetic

    c. both **a** and **b**

_____ related to electrons   **d.** neither **a** nor **b**

_____ determined by amount of charge

---

20. How could hitting a temporary magnet cause it to lose its magnetic properties?

_____

_____

_____

21. Explain how a spark can be released when you walk on a rug and then touch a doorknob.

_____

_____

_____

_____

_____

22. Distinguish between a step-down transformer and a step-up transformer in terms of voltage.

_____

_____

_____

23. List two of the four things you must have in order to use electricity.

_____

_____

_____

_____

24. How much current (in amps) is needed to light a bulb that has 0.5 Ω of resistance using a 1.5 V battery? (Show your work.)

_____

25. What is the voltage of the current leaving the transformer in the situation described below? Show your work.

turns in the primary coil (T_P) = 120

turns in the secondary coil (T_S) = 480

voltage in the primary coil (E_P) = 40 V

_____

_____

Estimado padre/madre de familia,

En las próximas semanas, su hijo o hija examinará lo que es la electricidad. Se investigarán diferentes fuentes de electricidad, lo mismo que el modo cómo las corrientes eléctricas fluyen a través de los circuitos. Se explorará también la relación entre la electricidad y el magnetismo. Cuando los estudiantes hayan terminado la Unidad 5, deberán ser capaces de dar respuesta a las siguientes preguntas, para captar las "grandes ideas" de la unidad.

1. ¿Cómo se puede producir la electricidad? (Cap. 13)

2. ¿Qué diferencia hay entre una pila y una batería? (Cap. 13)

3. ¿Cuáles son las partes que tienen en común todas las pilas químicas, y cómo funcionan? (Cap. 13)

4. ¿En qué se diferencian la corriente AC y la DC? (Cap. 14)

5. ¿Qué factores afectan la medida de las corrientes en los circuitos? (Cap. 15)

6. ¿Cuáles son algunas de las diferencias entre los circuitos en serie y los paralelos? (Cap. 15)

7. ¿Qué son los electroimanes? ¿Cómo se puede aumentar su fuerza? (Cap. 15)

8. ¿En qué se parece el flujo del agua al de la electricidad? (Cap. 15)

9. ¿Con qué unidades se mide la electricidad? ¿Cómo se relacionan estas unidades? (Cap. 15)

A continuación se menciona una actividad que, si usted quiere, puede practicar con su hijo o hija en la casa. Haga que su hijo o hija lleve una brújula alrededor de la casa o del apartamento y la ponga cerca de distintos aparatos. Dígale que se fije en cualquier movimiento de la aguja de la brújula en diversas posiciones, alrededor de cada aparato. Su hijo o hija debe hacer la prueba tanto con aparatos que estén funcionando como con los que estén apagados. Pídale que ponga la brújula tan cerca de los motores como sea posible sin peligro. (La aguja de la brújula debe tener una desviación cuando se la acerque a un motor eléctrico encendido.) Pídale a su hijo o hija que analice lo que observe. Comente las posibles causas. (La idea básica es que la aguja tendrá una desviación cerca de un motor encendido, porque se crea un campo magnético cuando el motor convierte la electricidad en movimiento.) Esta actividad simple va a ayudar a los estudiantes a ver la relación entre electricidad y magnetismo. Los estudiantes aprenderán más sobre la relación entre la electricidad y el magnetismo al ir avanzando en la unidad.

Atentamente,

Los materiales que aparecen abajo van a ser usados en clase para varias exploraciones de ciencia de la Unidad 5. Su contribución de materiales va a ser muy apreciada. He marcado algunos de los materiales en la lista. Si usted los tiene y quiere donarlos, por favor mándelos a la escuela con su hijo o hija para el

_____.

○ papel de aluminio
○ cartón (hojas y tubos)
○ perchas para ropa (de metal)
○ pinzas para colgar ropa
○ brújulas (magnéticas)
○ tiras de cobre
○ corchos
○ vasos (de papel; pequeños)
○ pilas D
○ franela
○ focos de linterna de mano
○ luces fluorescentes o de neón
○ pinzas
○ cerraduras de hierro
○ agujas de tejer (cortas)
○ guantes, tipo "latex"
○ lamparillas
○ imanes
○ cinta, tipo "masking"
○ plastilina para modelar
○ periódicos

○ clips para papel
○ tiras de plástico o reglas de plástico
○ envoltura de plástico
○ elásticos
○ sal
○ papel de lija
○ tornillos
○ células solares
○ lana de acero
○ alfileres
○ hilo
○ tachuelas o chinches
○ tiras de material vinílico, de aprox. 10 cm
○ arandelas
○ trigo soplado (cereal)
○ cable de cobre, hierro y níquel o cromo—aislado y sin aislar
○ bloques de madera (pequeños)
○ tarugos de madera
○ tiras de zinc

Desde ya, le agradecemos su ayuda.

En la Unidad 5, Sistemas electromagnéticos, explorarás lo que es la electricidad. Investigarás diversas fuentes de electricidad y el modo cómo las corrientes eléctricas fluyen a través de los circuitos. También vas a examinar la relación entre la electricidad y el magnetismo. Al ir leyendo la unidad, trata de responder a las siguientes preguntas. Estas son las "grandes ideas" de la unidad. Cuando puedas contestar estas preguntas, habrás logrado entender bien los principales conceptos de esta unidad.

1. ¿Cómo se puede producir la electricidad? (Cap. 13)

2. ¿Qué diferencia hay entre una pila y una batería? (Cap. 13)

3. ¿Cuáles son las partes que tienen en común todas las pilas químicas, y cómo funcionan? (Cap. 13)

4. ¿En qué se diferencian la corriente AC y la DC? (Cap. 14)

5. ¿Qué factores afectan la medida de las corrientes en los circuitos? (Cap. 15)

6. ¿Cuáles son algunas de las diferencias entre los circuitos en serie y los paralelos? (Cap. 15)

7. ¿Qué son los electroimanes? ¿Cómo se puede aumentar su fuerza? (Cap. 15)

8. ¿En qué se parece el flujo del agua al de la electricidad? (Cap. 15)

9. ¿Con qué unidades se mide la electricidad? ¿Cómo se relacionan estas unidades? (Cap. 15)

Spanish

Vocabulario

Alternating current (320)   **Corriente alterna** corriente eléctrica que revierte su dirección a intervalos regulares

Alternator (S97)   **Alternador** un aparato mecánico para generar corriente alterna

Ampere (amp) (343)   **Ampère** unidad básica para medir la corriente eléctrica, igual al coulomb de carga que pasa por un punto dado en un segundo

Battery (310)   **Batería** un grupo de pilas químicas conectadas que convierten energía química a energía eléctrica para producir corriente eléctrica

Chemical cell (305)   **Pila química** un aparato, que consiste de un electrolito y dos electrodos, que convierte la energía química a energía eléctrica para producir corriente eléctrica

Circuit (296, S101)   **Circuito** un pasaje continuo para el flujo de electrones

Conductor (301)   **Conductor** un material a través del cual fluye con facilidad la corriente eléctrica

Coulomb (343)   **Coulomb** unidad de medida de carga eléctrica, igual a la carga de 6.24×1018 (6.24 millones de millones) de electrones

Current (343, S88)   **Corriente** flujo de carga eléctrica a través de un conductor. La corriente se mide en ampères.

Domains (S94)   **Dominios** en materiales magnéticos, grupos de átomos en los cuales los campos magnéticos de la mayoría de los átomos están alineados en la misma dirección

Dry cell (310)   **Pila seca** un tipo de pila química en la cual el electrolito es una pasta más que un líquido

Electric current (300, 303, S86)   **Corriente eléctrica** el flujo de electrones de un lugar a otro

Electric field (S85)   **Campo eléctrico** la región de espacio alrededor de un objeto cargado eléctricamente en el cual se pueden observar los efectos de la fuerza eléctrica

Electric force (S84)   **Fuerza eléctrica** la fuerza que causa que dos objetos con cargas similares se repelan o que dos objetos con cargas desiguales se atraigan

Electricity (293)   **Electricidad** la energía de partículas negativas (electrones) que fluye en un conductor

Electrode (305)   **Electrodo** terminal que conduce electrones hacia o del electrolito en una batería

Electrolyte (305)   **Electrolito** sustancia que conduce electricidad; usualmente se disuelve en agua o en algún otro solvente

Electromagnet (338)　**Electroimán** imán producido por una corriente eléctrica que fluye a través de un resorte de alambre aislado, enroscado alrededor de un núcleo de acero o hierro

Electron (302)　**Electrón** una partícula, con carga negativa, que se encuentra en los átomos

Fuse (349)　**Fusible** aparato eléctrico que contiene un trozo de metal que se funde cuando demasiada corriente pasa a través del circuito que contiene el aparato

Galvanometer (306)　**Galvanómetro** instrumento que mide pequeñas cantidades de corriente eléctrica

Hertz (Hz) (321)　**Herzio** la unidad básica de medida de frecuencia, igual a 1 ciclo por segundo

Insulator (301)　**Aislante** material a través del cual pasa la corriente con dificultad, o a través del cual no pasa

Magnetic field (S92)　**Campo magnético** región del espacio alrededor de un imán en la cual se notan las fuerzas magnéticas

Magnetic force (317, S92)　**Fuerza magnética** la atracción o repulsión entre los polos de un imán; la fuerza que las corrientes eléctricas ejercen una sobre otra

Magnetic lines of force (317)　**Líneas de fuerza magnética** pasajes curvos, invisibles, entre los polos de un imán en los cuales el imán ejerce su fuerza

Motor (S100)　**Motor** aparato para convertir energía eléctrica en energía mecánica

Negative electrode (313)　**Electrodo negativo** la terminal que conduce los electrones hacia el electrolito en una pila química

Parallel circuit (331)　**Circuito paralelo** un circuito eléctrico arreglado de tal manera que la corriente pasa a través de más de un pasaje simultáneamente

Piezoelectric effect (323)　**Efecto piezoeléctrico** la producción de corrientes eléctricas por ciertos cristales cuando se los aprieta o se los estira

Positive electrode (313)　**Electrodo positivo** la terminal que conduce electrones lejos del electrolito en una pila química

Potential difference (S87)　**Diferencia potencial** la cantidad de trabajo ejecutado para mover una carga entre dos puntos en un campo eléctrico

Proton (302)　**Protón** una partícula cargada positivamente que se encuentra en el núcleo de un átomo

Resistance (329, S89)　**Resistencia** la oposición al flujo de la corriente en un circuito eléctrico

Resistor (330)　**Reóstato** aparato puesto en un circuito eléctrico para proporcionar resistencia

Series circuit (331)　**Circuito en serie** un circuito arreglado de tal manera que los compuestos se unen en su fin, en donde hay un solo pasaje para que la corriente fluya

Spanish

Static electricity (303, S84) **Electricidad estática** energía potencial en forma de una carga eléctrica estacionaria

Switch (336) **Interruptor** aparato para abrir y cerrar un circuito para empezar o detener el flujo de la corriente eléctrica

Variable resistor (330) **Reóstato variable** aparato en el cual la cantidad de resistencia puede ser cambiada para controlar el fluir de la corriente en un circuito eléctrico

Volt (V) (344) **Voltio** la unidad básica de medida de voltaje

Voltage (312, 344) **Voltaje** medida de la fuerza que empuja la corriente a través de un conductor, expresada en voltios

Watt (345) **Vatio** unidad de medida de potencia, igual a 1 joule por segundo

Wet cell (311) **Pila húmeda** tipo de pila química que usa un electrolito líquido

~~~ Chapter 13

Chapter 13
Exploration Worksheet

EXPLORATION 1

Electricity Working for You, page 296

Your goal	to use electricity to produce other forms of energy for use

Safety Alert!

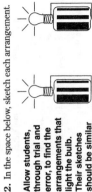

Illustration also on page 296 of your textbook.

You have probably seen warnings like the one at right on appliances or power tools. As useful as it is, electricity in large amounts is deadly and must be carefully controlled.

The familiar situations that you have been analyzing in the Case Studies on pages 294 and 295 of your textbook involve complex electrical parts and arrangements. Large quantities of electricity are used in the devices in Case Studies A and B. This is true for the operation of most appliances in homes, stores, or industry. A moderate amount is used in the devices in Case Studies C and D. Care must be exercised in using these amounts of electricity.

Most of the Explorations in this unit require only a small amount of electricity. They are quite safe to do. Here are four experiments in which a small amount of electricity works for you by producing other forms of energy. Watch for these energy forms. In these experiments the electricity is supplied by dry cells such as those used in flashlights.

Activity 1: Shedding a Little Light

You Will Need

- a length of magnet wire
- sandpaper
- a flashlight bulb
- a D-cell
- a rubber band

What to Do

1. Begin by sanding the enamel off the last few centimeters of each end of the wire to expose the copper underneath. Then, using only the other items listed, find all of the different arrangements that will light the bulb.

2. In the space below, sketch each arrangement.

Allow students, through trial and error, to find the arrangements that light the bulb. Their sketches should be similar to the following:

HRW material copyrighted under notice appearing earlier in this work.

Photo/Art Credits

Abbreviated as follows: (t) top; (b) bottom; (l) left; (r) right; (c) center; (bkgd) background.

Photo Credits
Front cover: (bkgd), Page Overtures; (bl), Randy Gates/Morgan-Cain & Associates. Back cover: (bkgd), Page Overtures; (bl), Jeff Smith/FotoSmith/Reptile Solutions of Tucson. Title page (t): (bkgd), Page Overtures; (bl), Jeff Smith/FotoSmith/Reptile Solutions of Tucson. Page 6(c), Scott Van Osdol/HRW; 21(br), Sam Dudgeon/HRW; 37(br), John Langford/HRW; 39(bl), Scott Van Osdol/HRW; 79(t), David Phillips/HRW; 91(bl), Scott Van Osdol/HRW; 98(br), Sam Dudgeon/HRW; 105(cr), John Langford/HRW; 106(br), Scott Van Osdol/HRW; 123(bl), David Phillips/HRW.

Art Credits
All work, unless otherwise noted, contributed by Holt, Rinehart and Winston
Page 8, Dodson Publication Services; 9, Stephen Durke/Washington-Artists' Represents, Inc.; 11, Stephen Durke/Washington-Artists' Represents, Inc.; 13, Stephen Durke/Washington-Artists' Represents, Inc.; 14, Richard Wehrman; 16, Dodson Publication Services; 17, David Griffin; 18, Patti Bonham/Washington-Artists' Represents, Inc.; 23–24, Jim Pfeffer; 28, Mark Persyn; 32, David Fischer; 34, David Merrell/Suzanne Craig Represents; 35, Morgan-Cain & Associates; 36, Dodson Publication Services; 41, Patti Bonham/Washington-Artists' Represents, Inc.; 45, Morgan-Cain & Associates; 48–50, Morgan-Cain & Associates; 51, Uhl Studio; 56, Blake Thornton/Rita Marie & Friends; 57, Morgan-Cain & Associates; 58, Morgan-Cain & Associates; 59, Morgan-Cain & Associates; 60, Morgan-Cain & Associates; 67–70, Morgan-Cain & Associates; 72, Morgan-Cain & Associates; 92, Stephen Durke/Washington-Artists' Represents, Inc.; 93, Stephen Durke/Washington-Artists' Represents, Inc.; 94(t), Richard Wehrman; 94(b), Stephen Durke/Washington-Artists' Represents, Inc.; 95, Dodson Publication Services; 97, Morgan-Cain & Associates; 99, Jim Pfeffer; 101, Mark Persyn; 102, David Fischer; 103, David Merrell/Suzanne Craig Represents; 104, Morgan-Cain & Associates; 106, Morgan-Cain & Associates; 107, Patti Bonham/Washington-Artists' Represents, Inc.; 109, Morgan-Cain & Associates; 110(t), Uhl Studio; 110(b), Morgan-Cain & Associates; 113(t), Morgan-Cain & Associates; 113(b), Blake Thornton/Rita Marie & Friends; 114, Morgan-Cain & Associates; 115, Morgan-Cain & Associates; 118, Morgan-Cain & Associates; 119, Morgan-Cain & Associates; 120, Morgan-Cain & Associates.

HRW material copyrighted under notice appearing earlier in this work.

Answer Keys

Unit 5: Electromagnetic Systems

Contents

Name _____ Date _____ Class _____

Exploration 1 Worksheet, continued

Repeat with a rolled-up length of aluminum foil and then with a length of nichrome wire. Record your observations here.

The aluminum foil and nichrome wire become very hot. Number 32 nichrome (an alloy of nickel and chromium) will glow with the electrical energy supplied by a single D-cell.

Bring your hand close to each piece being tested, but do not touch any of them. What do you feel?

Students should feel some heat radiating from the steel wool, the foil, and the nichrome wire.

5. What form of energy does the electricity produce?

   Heat energy

6. What are some examples of devices in which this type of electrical energy is used?

   Heating elements in stoves, heaters, toasters, and toaster ovens; light bulb filaments; and car windshield defrosters

7. Do Activities 1 and 2 demonstrate the same principle? Explain.

   Yes; Activities 1 and 2 both demonstrate an electric circuit because electricity is flowing in a circuit through wires connected to a D-cell. The steel wool, aluminum foil, and nichrome wire from Activity 2 are similar to the filament in the flashlight bulb from Activity 1.

As you may have guessed from the title of this unit, there is a connection between electricity and magnetism. The following Activities will help illustrate that connection.

SCIENCEPLUS • LEVEL BLUE   5

Name _____ Date _____ Class _____

Exploration 1 Worksheet, continued

3. What form(s) of energy does the electricity produce?

   Light energy (bulb) and heat energy (bulb and wire)

4. What are other examples of electricity being used in this way?

   Electric signs, glowing heat elements, street lights, automobile lights, flashlights, and stadium lights

5. You have constructed an electric *circuit*. What parts make up this circuit?

   The electric circuit is made up of a D-cell, wire, and a flashlight bulb.

   Check the dictionary to find out the origin of the word *circuit*. How is it significant?

   Sample answer: The term *circuit* comes from a Latin word that means "to go around." It is an appropriate term because electricity follows, or goes around, a circuit.

   What would you say an electric circuit is?

   Sample definition of an electric circuit: a loop through which electricity can move in a certain direction

Activity 2: The Heat Is On

You Will Need

- two 30 cm lengths of magnet wire
- sandpaper
- modeling clay
- a clothespin
- 2 D-cells
- a wide rubber band
- a strand of steel wool
- aluminum foil
- a thin nichrome wire (10 cm long)

What to Do

1. Sand the enamel off of the last 5 cm of the ends of each length of magnet wire. Then make a small loop at one end of each wire.

2. Bend the wires and support them with modeling clay, as shown on page 297 of your textbook.

3. Attach the ends of the wires to the D-cell, securing them with a wide rubber band.

4. Place a strand of steel wool through the loops and let it rest in contact with each loop. What do you observe? **(NOTE: Do not leave the test materials in contact with the loops for very long because this will quickly drain the cell of its electrical energy.) The strand of steel wool completely or partially melts.**

4   UNIT 5 • ELECTROMAGNETIC SYSTEMS

Activity 4: Let's Get Moving!

You Will Need

- 2 D-cells
- a paper clip
- 2 lengths of magnet wire (10 cm and 250 cm long)
- a bar magnet
- a support stand with ring clamp
- a cork
- a wide rubber band
- a narrow strip of cardboard (2 cm × 30 cm)
- 2 thumbtacks or screws
- a wood block
- sandpaper

What to Do

1. Sand the enamel off of the last 2 or 3 cm of the ends of each wire. Then assemble the circuit as shown on page 299 of your textbook.

2. Have one person hold a strong magnet near the cork while the other person presses the contact switch to complete the circuit. Observe what happens. **Caution: Don't keep the switch closed for very long.**

   The coil of wire is repelled by or attracted to the magnet, depending on which end of the bar magnet is held near the coil.

   Open the switch. What happens?

   The coil of wire returns to its original position. By opening and closing the switch, the coil of wire can be made to swing back and forth.

3. Using the other end of the magnet, repeat step 2. What happens?

   When the switch is closed, the coil is attracted to or repelled by the magnet, depending on which end of the bar magnet was originally held near the coil. When the switch is open, the coil moves back to its original position. This is the reverse of what happened in step 2.

4. What kind of energy is produced by the electricity?

   The electricity is producing magnetic energy, which in turn produces energy of motion, or kinetic energy.

5. What are some practical examples of how energy works for you in this way?

   Sample answer: Doorbells and electric motors use the magnetic field associated with electricity to create movement.

A Home Project

Make a battery tester. Use one of the arrangements you discovered in Activity 1 of Exploration 1. Devise a tester that consists of a light bulb with two wires connected to it. Touch the ends of the wire to the battery. The brightness of the light bulb will indicate the strength of the battery.

Activity 3: The Electricity-Magnetism Connection

You Will Need

- a D-cell
- a compass
- 2 thumbtacks or screws
- a wood block
- a paper clip
- 2 lengths of magnet wire (15 cm and 25 cm long)
- a rubber band
- sandpaper

What to Do

1. Sand the enamel off of the last 2 or 3 cm of the ends of each wire. Then set up the apparatus as shown below. Align the compass needle, and place the wire over the compass in a north-south direction so that the wire lines up with the compass needle.

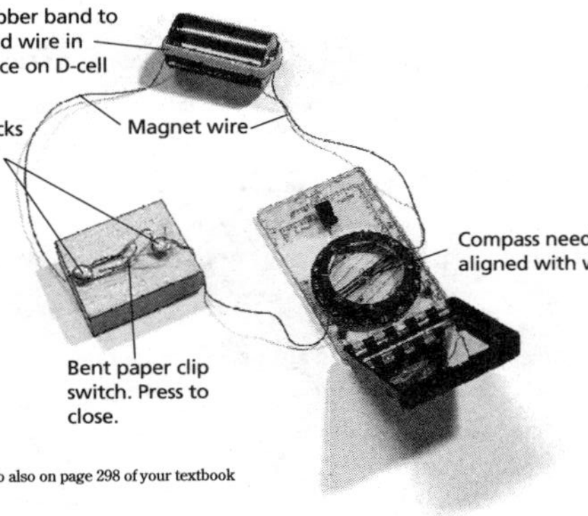

Rubber band to hold wire in place on D-cell

Magnet wire

Thumbtacks or screws

Compass needle aligned with wire

Bent paper clip switch. Press to close.

Photo also on page 298 of your textbook

2. Close the electrical circuit by pressing the contact switch. What happens? **Caution: Don't keep the switch closed for very long.**

   The compass needle is deflected to the right or left, depending on which poles of the D-cell the wires are attached to.

3. What kind of energy does the electricity produce in this experiment?

   Magnetic energy and heat energy (from the wire)

4. Can you think of any everyday applications that make use of electricity in this way?

   Answers will vary. Sample answer: Speedometers, odometers, and fuel gauges make use of electricity in this way.

Answers • Chapter 13

Name _____ Date _____ Class _____

A Theory of Charged Particles, page 301

Scientists explain events such as those seen in A Classical Current Demonstration on page 300 of your textbook in this way:

1. When one material is rubbed against another, friction causes charged particles to move from one material to the other. The accumulated charge is indicated by the charged material's ability to attract lightweight or finely powdered substances. When did this happen in the previous demonstration?

This occurred when the plastic strip rubbed with plastic wrap and the vinyl strip rubbed

with flannel both caused the wheat puff to move toward the nail. _____

_____

2. There are two kinds of charged particles: positively charged particles and negatively charged particles. Why might you conclude that the charges on the plastic strip are different from those on the vinyl strip?

When the plastic strip gave its charge to the nail, and the nail passed the charge on to the

wheat puff, the wheat puff moved away from the nail. When the vinyl strip gave its charge to

the nail, the wheat puff was attracted to the nail. This demonstrates that the charge on the

vinyl strip must be opposite that on the plastic strip. _____

3. Objects that have the same kind of charge tend to repel one another. Objects that have different charges tend to attract one another. When did you observe these effects in the demonstration?

After the wheat puff made contact with the nail, they were both negatively charged, so the

wheat puff was repelled. Then the nail was positively charged by the vinyl strip, so the

wheat puff was attracted. _____

4. Charged particles pass easily through certain materials, called **conductors**, but pass with difficulty or not at all through other materials, called **insulators** (nonconductors). What evidence is there that iron is a conductor of electricity?

Iron must be a conductor of electricity because the charge seemed to travel through the

iron nail. _____

Name _____ Date _____ Class _____

A Classical Current Demonstration, page 300

Here is a demonstration similar to one first done hundreds of years ago. Make a setup like the one shown below, and try it yourself. Follow the steps closely.

You Will Need

- a plastic strip, such as a plastic ruler
- plastic wrap
- scissors
- 30 cm of thread
- a wheat puff
- a wire clothes hanger
- a nail
- a paper cup
- a vinyl strip
- flannel cloth

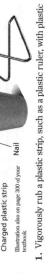

Coat hanger
Thread
Notched paper cup
Wheat puff
Nail

Charged plastic strip

Illustration also on page 300 of your textbook

1. Vigorously rub a plastic strip, such as a plastic ruler, with plastic wrap.

2. Quickly touch the strip or ruler to the end of the nail. What happens to the wheat puff?

The wheat puff moves toward the pointed end of the nail and then

"jumps" away from it. _____

3. Rub the plastic strip again and bring it close to the end of the nail without quite touching it. Observe the wheat puff.

The wheat puff moves away from the nail. _____

4. Repeat the experiment using a vinyl strip rubbed vigorously with flannel cloth. What happens to the wheat puff now?

The wheat puff moves toward the nail. _____

How do you explain the events in this demonstration?

When the electrically charged plastic strip touches the nail, it transfers

its charge to the nail. The nail attracts the uncharged wheat puff and

transfers its charge to the wheat puff. After the wheat puff becomes

charged, it is repelled by the nail because they are similarly charged.

The wheat puff is attracted to the nail when it is charged with the vinyl

strip because the two are oppositely charged. _____

Chapter 13
Exploration Worksheet

EXPLORATION 2

Constructing a Current Detector, page 306

| Your goal | to build a galvanometer, an instrument used to detect and measure electrical current |

Your task is to design and construct a homemade galvanometer.

You Will Need

- a small magnetic compass
- insulated wire
- anything else you need to hold the parts in place
- paper clip
- magnet wire
- sandpaper
- paper cup
- concentrated lemon juice

Hints

1. Remember the results of Exploration 1, Activity 3 on page 298 of your textbook.

2. Try coiling the insulated wire around the compass. Use different numbers of turns and observe the effect. Record your observations.

   Observations will vary. Students should discover that the more turns there are in the coil, the stronger the coil's magnetic force becomes. As they experiment with their galvanometers, students should also realize that the more turns of wire there are, the more sensitive their galvanometers become.

3. Leave the two ends of the wire free so that you can attach them to the source of the small current.

4. Position the galvanometer so that the compass needle and the coil of wire are parallel to one another.

5. Test your galvanometer using a small current, which can be obtained from a lemon-juice cell. To make a lemon-juice cell, place a straightened paper clip and a piece of sanded magnet wire into a small cup of concentrated lemon juice. The paper clip and the wire are the electrodes, and the lemon juice is the electrolyte. Hook the free ends of the galvanometer wire to the electrodes. What happens?

   The compass needle is deflected, indicating that an electric current is flowing.

6. Will your galvanometer be able to give you any information about the size of the current? Explain.

   Yes. The stronger the electric current is, the farther the compass needle is deflected.

A Theory of Charged Particles, continued

You might experiment by replacing the nail in the demonstration with a glass rod, a wooden stick, a copper wire, or objects made of other materials. Which are conductors? Which are insulators?

Wood and glass do not conduct electricity and are therefore insulators. Copper wire is a very good conductor.

The drawings below apply the theory just stated to the previous demonstration. Express in your own words what is taking place in each drawing.

A   Plastic wrap   Plastic strip

B   Flannel   Vinyl strip

Illustration also on page 301 of your textbook

Students should reword their answers to A Classical Current Demonstration (on page 300 of the textbook) to demonstrate an understanding of the transfer of charged particles through a conductor.

≋ **Answers • Chapter 13**

Chapter 13
Review Worksheet

Challenge Your Thinking, page 307

1. All Charged Up

a. One by one, a negatively charged plastic ruler is brought near three light, foil-covered spheres suspended by nylon threads. The ruler repels sphere A and attracts spheres B and C. Sphere A attracts sphere B, and sphere C attracts sphere B. Do you have enough information to determine the charges on spheres B and C? Why or why not?

Illustration also on page 307 of your textbook

No. Although sphere A must be negatively

charged because it is repelled by the nega-

tively charged ruler, spheres B and C could

be either uncharged or positively charged. If

sphere B is uncharged, then sphere C must

be positively charged, and vice versa.

b. Jeff rubbed two pieces of plastic wrap with a sock and then suspended them, as shown on page 307 of your textbook. Note what he observed.

The two pieces of plastic wrap moved away from each other.

c. Jeff then brought one of the pieces near a table. Again observe what he saw.

The piece of plastic wrap moved toward the table.

d. Explain what is happening.

When the two pieces of plastic wrap are rubbed with the sock, they

acquire the same kind of electrical charge, so they repel each other.

Because of polarization, there is an attraction between the charged

plastic wrap and the uncharged table. The plastic wrap has very little

mass, so it is pulled toward the more massive table.

Try Jeff's experiment yourself.

Chapter 13 Review Worksheet, continued

2. Sure Shot

When spray painting a screen, the screen is given an electric charge. Why doesn't the paint spray go through or around the screen?

Illustration also on page 307 of your textbook

A charged object attracts oppositely charged objects. A

charged object can also attract an uncharged object by pol-

arizing the charges in the uncharged object. The paint spray

does not go through or around the screen because the

charged screen attracts the less massive, uncharged paint

particles in the spray.

3. Current Puzzle

Complete the following statements using the puzzle at left. Locate the answers in the puzzle by searching horizontally, vertically, or in a combination of both directions. Cross out the letters of each answer in the puzzle. The letters that remain will tell you something about an electric current.

a. A material that allows charges to go through it is a(n) **conductor** .

b. A material that does not allow charges to go through it is a(n) **insulator** .

c. J.J. Thomson discovered a charged particle that moves readily; it is called a(n) **electron** .

d. This type of particle has a(n) **negative** charge.

e. The other charged particle in materials, which is more massive and does not move, is called a(n) **proton** and has a(n) **positive** charge.

f. If a material has equal numbers of these two kinds of particles, the material is **uncharged** .

Mystery statement: **Current = charges in motion**

4. That's a Wrap

Some kinds of plastic wrap can be stretched tightly over a container and down its sides. The plastic wrap sticks to the sides of the container. Why?

The stretching of the wrap causes the wrap to become charged. The attrac-

tive forces between the charged wrap and the uncharged container cause

the wrap (which has a very small mass) to adhere to the container.

Name _____ Date _____ Class _____

Chapter 13 Assessment, continued

4. Circle the materials that are good conductors of electricity.

   a. glass   b. wood

   c. (iron)   d. (zinc)

   e. (acid)   f. plastic

5. Use what you've learned about static electricity to explain the illustration below.

CHALLENGE

Illustration for Interpretation

Sample answer: The comb becomes negatively charged as it picks up electrons from the girl's hair. Her hair is now positively charged, so it is attracted to the negatively charged comb and moves toward it.

6. Name at least one way your life would be different if plastic were not a good insulator.

Answers will vary but should reflect the fact that plastic is used to cover wiring, as on appliance cords and other items that carry electricity.

CHALLENGE

Short Response

16   UNIT 5 • ELECTROMAGNETIC SYSTEMS

Name _____ Date _____ Class _____

Chapter 13
Assessment

Word Usage

1. Write one or two sentences to explain the difference between the words *electrolyte* and *electrode*.

   An *electrode* is the positive or negative component of a cell. It conducts electrons into or out of the *electrolyte*, a solution of chemicals that conducts electrons between the positive and negative electrodes. The conducting path causes an electric current to flow.

Correction/ Completion

2. The following sentences are incorrect or incomplete. Your challenge is to make them correct and complete.

   a. When one material is rubbed against another, electricity causes charged particles to move from one material to the other.
   When one material is rubbed against another, *friction* causes charged particles to move from one material to the other.

   b. An electrostatic charge involves the continuous flow of electrons through a material such as a copper wire.
   An *electric current* involves the continuous flow of electrons through a material such as a copper wire.

Short Responses

3. Identify each of these setups with a single term.

   a. A free-floating magnetic needle used to tell direction   **Compass**

   b. A group of connected cells   **Battery**

   c. A wire coiled around a compass to indicate an electric current   **Galvanometer**

SCIENCEPLUS • LEVEL BLUE   15

Chapter 14
Exploration Worksheet

EXPLORATION 1

Chemical Cells, page 310

Your goal	to make and test a wet cell and a dry cell

Safety Alert!
Wear goggles and latex gloves when working with ammonium chloride.

Experiment 1: Dry Cells

You Will Need
- a homemade or commercial galvanometer
- a 40 cm length of magnet wire
- masking tape
- rubber bands
- 2 zinc strips (3 cm × 8 cm)
- 2 copper strips (3 cm × 8 cm)
- blotting paper or filter paper
- ammonium chloride solution
- a container to hold the ammonium chloride solution
- forceps
- latex gloves

What to Do
1. Make a single-cell sandwich like the one shown below at left.
2. Measure the amount of deflection this sandwich produces in your galvanometer.
3. Now make a double-cell sandwich, as shown in the second illustration below.
4. Measure the deflection it produces in the galvanometer. How does it compare with the deflection caused by the single-cell sandwich?

The deflection for the double-cell sandwich should be about twice as much as that for the single-cell sandwich.

Because the double-cell sandwich consists of more than a single cell, it is known as a **battery**.

Single-Cell Sandwich
Rubber bands
Tape
Copper (becomes positively charged)
Blotting paper soaked in ammonium chloride solution
Zinc (becomes negatively charged)
Homemade galvanometer
Magnet wire
Plastic-foam cup

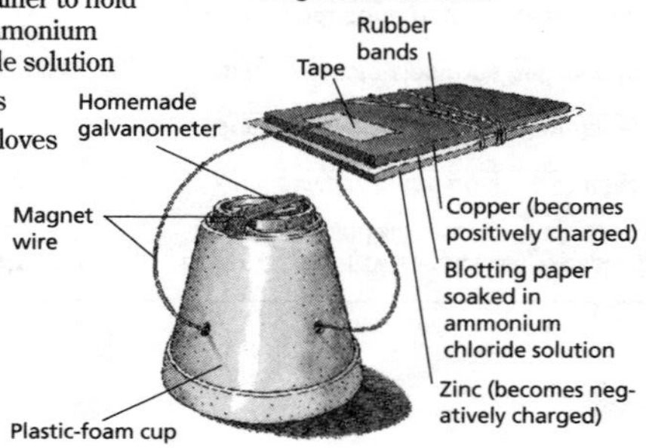

Double-Cell Sandwich
Zinc (–)
Copper (+)
Soaked blotting paper

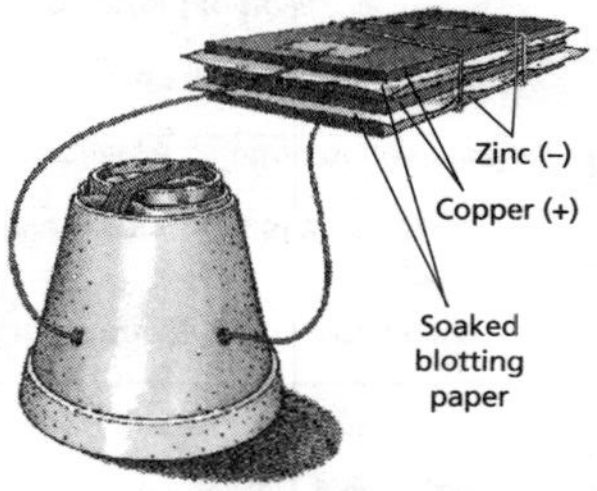

Illustrations also on page 310 of your textbook

SCIENCEPLUS • LEVEL BLUE   17

Exploration 1 Worksheet, continued

Questions

1. What accounts for the difference in the galvanometer readings for the single-cell and double-cell sandwiches?
The amount of chemicals available for conversion from chemical energy to electrical energy is double.

What would be the effect of adding more layers to the sandwich?
The current would increase even more.

2. Which electrodes were linked together in converting a single-cell sandwich to a double-cell sandwich?
Copper (+) and zinc (–) are placed together when the double-cell sandwich is formed.

Experiment 2: Wet Cells

You Will Need
- a commercial galvanometer
- a zinc strip (2 cm × 15 cm)
- a copper strip (2 cm × 15 cm)
- salt solution
- a 250 mL beaker
- 2 pieces of magnet wire (each 20 cm long)
- 2 alligator clips or clothespins

What to Do
1. Make a setup like that shown below.
2. Add enough salt solution to cover about half of the metal strips. Connect the strips to the galvanometer as shown.

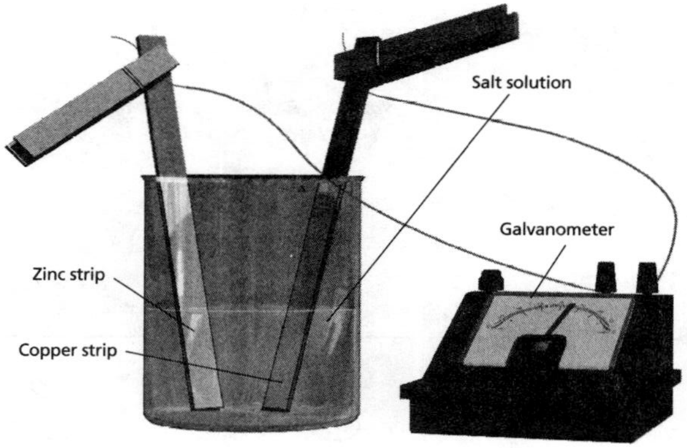

Galvanometer
Salt solution
Zinc strip
Copper strip

Illustration also on page 311 of your textbook

Name _____ Date _____ Class _____

Exploration 1 Worksheet, continued

3. Observe the galvanometer. Record the highest reading reached.

Draw a sketch showing your answer.

Zinc Copper   Zinc Copper

Name _____ Date _____ Class _____

Exploration 1 Worksheet, continued

3. Observe the galvanometer. Record the highest reading reached.

What happens to the reading?

The highest reading occurs when the salt solution is added to the beaker. However, the reading on the galvanometer decreases almost immediately, indicating that the current is weakening. As more salt solution is added, covering more of the electrodes, the current increases.

Observe each electrode carefully. What happens to each electrode?

A gas is produced and collects in small bubbles on the copper strip (electrode).

4. Add enough salt solution to fill the beaker.
5. Record the galvanometer reading once again.

Questions

1. How would you connect two wet cells to get more current?

Run one wire from the zinc electrode in one cell to the copper electrode in the other cell. Connect the output wires from the free electrodes in each cell to the galvanometer.

≋ Chapter 14

2. What is one way of increasing the current in a wet cell? How would you explain this?

Increase the amount of salt solution. This covers more of the electrodes and increases the number of electrons available in the solution.

3. What other factors might be altered in a wet cell to increase its current output?

Answers will vary. Possibilities include increasing the number of cells, increasing the size of the zinc and copper strips, and increasing the concentration of the salt solution.

4. What energy changes take place in the operation of dry and wet chemical cells?

Chemical energy is converted into electrical energy.

20   UNIT 5 • ELECTROMAGNETIC SYSTEMS

Name _____ Date _____ Class _____

Exploration 2 Worksheet, continued

2. The voltage of each cell on the previous page is marked. Notice how several cells of different size have the same voltage. How can that be?

Different-sized cells with the same voltage can be made by varying the amount and kinds of chemicals used in the cells.

What does *voltage* mean to you?

Answers may vary. Students will probably say that the voltage indicates the strength of a battery.

3. Look at the graph below and answer the questions that follow.

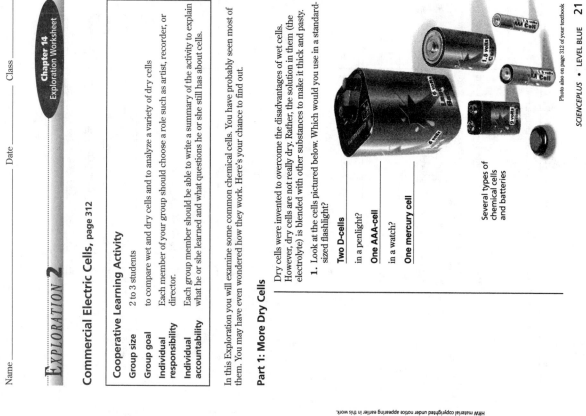

Relative Power of Dry Cells

Mercury cell

Alkaline cell

Ordinary cell

Strength of current

Number of hours (medium load)

Graph also on page 313 of your textbook

Which type of cell gradually "winds down"?

The alkaline cell

Which type loses power quickly?

The ordinary cell

Which type of cell would you probably use to power devices that require a steady current?

The mercury cell

4. Use the descriptions on the next page to help you label the parts of the batteries shown on this page and the next.

An example of labeling can be found on page 313 of the Annotated Teacher's Edition.

6 V

9 V

D

Illustration also on page 313 of your textbook

22 UNIT 5 • ELECTROMAGNETIC SYSTEMS

Name _____ Date _____ Class _____

EXPLORATION 2

Commercial Electric Cells, page 312

Cooperative Learning Activity

Group size	2 to 3 students
Group goal	to compare wet and dry cells and to analyze a variety of dry cells
Individual responsibility	Each member of your group should choose a role such as artist, recorder, or director.
Individual accountability	Each group member should be able to write a summary of the activity to explain what he or she learned and what questions he or she still has about cells.

In this Exploration you will examine some common chemical cells. You have probably seen most of them. You may have even wondered how they work. Here's your chance to find out.

Part 1: More Dry Cells

Dry cells were invented to overcome the disadvantages of wet cells. However, dry cells are not really dry. Rather, the solution in them (the electrolyte) is blended with other substances to make it thick and pasty.

1. Look at the cells pictured below. Which would you use in a standard-sized flashlight?

Two D-cells

in a penlight?

One AAA-cell

in a watch?

One mercury cell

Several types of chemical cells and batteries

Photo also on page 312 of your textbook

SCIENCEPLUS • LEVEL BLUE **21**

Name _____ Date _____ Class _____

Exploration 2 Worksheet, continued

Like all chemical cells, the ordinary dry cell has two electrodes, or conductors, and an electrolyte solution. The *positive electrode* has two parts—a *graphite rod* in the center of the cell and a mixture of *manganese oxide* and *powdered carbon* surrounding the graphite rod. The *negative electrode* is zinc; it makes up the sides and bottom of the cell. The *electrolyte* fills the space between the electrodes. It consists of *ammonium chloride* paste. At the top of the cell is an *insulator*. *Batteries consist of at least two individual cells joined together by conducting strips.*

In the mercury cell, the *positive electrode* consists of a small block of zinc. The *electrolyte* is *potassium oxide*. The *negative electrode* is a layer of *mercury oxide*. The *electrolyte* is *potassium hydroxide*.

5. The *alkaline cell* differs from an ordinary dry cell in two major ways. First, the negative electrode is made of spongy zinc. Second, the electrolyte is potassium hydroxide, a strong base. What effect do these differences have on the power output of the alkaline cell?

From the graph, students may infer that these differences make the alkaline cell last longer and provide a steadier current.

Part 2: Other Cells

1. A powerful surge of electric current is needed to crank an automobile engine. This surge of current is provided by a group of cells joined together in a battery. Study the drawing below to discover or infer the answers to the questions on the next page.

Mercury   AA   AAA

|—1 cm—|

Illustration also on page 313 of your textbook

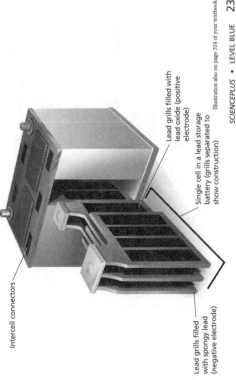

Intercell connectors

Single cell in a lead storage battery (grills separated to show construction)

Lead grills filled with spongy lead (negative electrode)

Lead grills filled with lead oxide (positive electrode)

Illustration also on page 314 of your textbook

≋ Chapter 14

Name _____ Date _____ Class _____

Exploration 2 Worksheet, continued

a. What substances make up (1) the two electrodes and (2) the electrolyte in a car battery?

The negative electrode is made of spongy lead; the positive electrode is made of lead oxide; the electrolyte is sulfuric acid.

b. Why is such a large battery needed for a car?

Because a large surge of electricity is needed to start the car's engine

2. Automobile batteries have a limited life span, and not all batteries last the same amount of time. Why do batteries wear out?

Batteries wear out because the electrodes are worn away or damaged by the chemical reactions continually taking place in the battery.

Why do some wear out sooner than others?

Some batteries wear out quickly because they are made from less durable materials or because they are used more often.

3. Research "maintenance-free" batteries. How do they work?

Maintenance-free batteries use a sulfuric acid gel as an electrolyte rather than a sulfuric acid solution made with distilled water.

How are they different from standard batteries?

Maintenance-free batteries do not require the addition of distilled water because they are sealed to prevent evaporation of the electrolyte.

CONCENTRATED SULFURIC ACID CAUTION

Highly corrosive. Avoid spillage when pouring.

Illustration also on page 314 of your textbook

Answers • Chapter 14

EXPLORATION 3

Moving Magnets and Wire Coils, page 315

Cooperative Learning Activity

Group size 2 to 3 students

Group goal to demonstrate how a current can be created by the relative motion of magnets and wire coils

Individual responsibility Each member of your group should choose a role such as coil monitor, magnet mover, or meter reader.

Individual accountability Each group member should be able to present the group's experimental design for step 6(c) to the class.

Part 1: Building Your Own

You Will Need

- a commercial galvanometer
- a 150 cm length of magnet wire
- a cardboard tube
- a strong bar magnet
- sandpaper

What to Do

1. Sand the enamel off of the last 2 or 3 cm of the ends of the magnet wire. Wrap the magnet wire around the tube to make a coil as illustrated on page 315 of your textbook. Attach the bare ends of the wire to a commercial galvanometer.

2. While watching the galvanometer, move a bar magnet into the coil, hold it there for a moment, and then remove it. Is the galvanometer needle affected?

   Yes. The needle is deflected while the magnet is moving.

3. Repeat step 2 several times, moving the magnet at different speeds. What do you observe?

   The faster the magnet is moved, the more the needle is deflected.

   Does moving the magnet into the coil have a different effect on the galvanometer than moving the magnet out of the coil? Explain.

   Yes. Changing the direction that the magnet moves changes the

   direction that the needle is deflected.

   What might this suggest about the direction of current flow through the coil?

   The direction of the current has also reversed.

Exploration 2 Worksheet, continued

≋ **Chapter 14**

4. For many applications, the *nickel-cadmium* cell is replacing both lead-acid batteries and dry cells. Find out how this type of cell works.

   The nickel-cadmium cell works in a manner similar to a lead-acid cell,

   but it uses nickel oxide for the positive electrode and cadmium for the

   negative electrode. A potassium hydroxide solution is the electrolyte.

   _____

   _____

5. Unlike dry cells, nickel-cadmium cells and lead-acid batteries can be *recharged*. What does this mean?

   The ability to supply electricity can be restored after the cell or battery

   has run down.

   _____

   How is this property useful?

   This property is useful because the same battery can be used over

   and over again.

   _____

Name _____ Date _____ Class _____

Exploration 3 Worksheet, continued

4. Disconnect the galvanometer and move the magnet to see whether the magnet itself is affecting the galvanometer. Record your observations.

The magnet alone will not produce an effect.

5. Connect the galvanometer to the coil again. This time, hold the magnet still, but pass the coil over the magnet. What do you observe?

The needle is deflected.

6. Use your observations to answer the following questions:

a. How can a magnet help produce electricity?

A magnet can produce electricity by moving through a coil of wire.

b. How is the direction of the current affected by the motion of the magnet?

When the motion of the magnet is reversed, the direction of the electric current is also reversed.

c. How does the speed of the magnet's motion affect the amount of electricity generated?

When the speed of the magnet's motion increases, the amount of electricity that is produced also increases.

d. What changes in forms of energy occur in this investigation?

Chemical energy in the person moving the magnet is changed into kinetic energy as the magnet is moved back and forth. Moving the magnet changes the kinetic energy into electrical energy.

e. Could a stationary magnet ever produce electricity? Explain.

Yes, if a coil of wire surrounding the magnet moves

f. Would current still be generated in the wire if the wire were broken at some point? Why or why not?

Yes. Initially, current would be generated, but it would cease to flow at the point where the circuit is broken.

Chapter 14

Name _____ Date _____ Class _____

Exploration 3 Worksheet, continued

Part 2: Francesca's Experiment

Francesca devised an experiment to answer questions raised by Part 1 of this Exploration. She started with a wire coil of 15 turns. Illustrations (a) through (c) show the galvanometer readings she recorded. Then she used a coil with twice as many turns. Illustrations (d) and (e) show these readings. Francesca tried each part of this experiment three times and obtained similar results each time. What conclusions do you think she drew for each part? The illustrations below provide some hints.

Illustrations also on page 316 of your textbook

a.   b. Magnet moved inside coil   c. Magnet moved faster inside coil

d. Magnet moved at the same speed as in (c)   e. Magnet moved out of and away from coil at the same speed as in (d)

Analysis

1. When a magnet is moved inside a coil of wire, _____**an electric current**_____ is detected in the wire, which _____**changes**_____ its direction when the magnet is moved in the opposite direction inside the coil.

2. A larger current is produced if _____**the magnet moves more quickly**_____ or if _____**the number of coils of wire is increased**_____

3. Suppose Francesca moved the magnet into and out of the coil 15 times in a minute. What would happen to the current?

The current would alternate direction.

How many times per minute would the current go first in one direction and then in the opposite direction?

The current would move in one direction 15 times and in the opposite direction 15 times per minute.

Answers • Chapter 14

Exploration 3 Worksheet, continued

4. What do you think would happen if Francesca held the magnet stationary and moved the wire coil instead? Why?

An electric current would be produced because the magnet would still

be moving in and out of the coil.

Part 3: A Related Experiment

Francesca made an important discovery: When a magnet is moved through a coil of conducting wire, electricity is generated. Both the number of coils and the speed of movement of the magnet affect the amount of current produced. Francesca also discovered that the direction in which the magnet was moved made a difference. If the magnet was moved in one direction, current flowed one way. If the magnet was moved in the other direction, current flowed the other way. Let's examine the findings of a related experiment.

But before you begin, think a little bit about how a magnet exerts its influence. Does the magnet have to touch something to have an effect, or does its force act through space? Look at the photo at the upper right on page 317 of your textbook. It shows a magnet on which iron filings have been sprinkled. Do you see evidence that (a) the iron filings have been attracted and that (b) the *magnetic force* is exerted through space along curved paths? We call these paths *magnetic lines of force.*

Look at the series of illustrations on page 317 of your textbook, which represent the results of the experiment. The wire is being moved while the magnet is held stationary. The arrows between the north and south poles of the magnet represent the lines of magnetic force.

Analysis

1. How do the results compare to those of Francesca's experiment?

The results are similar. A current is produced whether the wire is

moving or the magnet is moving.

2. What happens when the wire is momentarily motionless as it changes direction, as in (a)?

No current is produced.

3. What happens when the wire is moved parallel to the magnetic lines of force, as in (d)?

No current is produced.

4. What role do the magnetic lines of force appear to play in the generation of electricity?

They appear to cause an electric current to flow in a wire when the

wire is crossing and not moving parallel to the lines of force.

Chapter 14

Challenge Your Thinking, page 324

1. Generator X

The sequence of pictures below shows a type of generator in action. Use the pictures to help you answer the questions that follow.

Gaps in copper ring

a. Not moving

b.

c.

d.

Illustrations also on page 324 of your textbook.

a. Examine the construction of this generator. How does this generator work? **As the wire loop turns through the magnetic field, a current is produced in the wire. The ends of the wire are connected to two different segments of the armature. As the current flows in one direction, this armature segment touches one contact. As the current flows in the other direction, the other armature segment touches the other contact.**

b. Study the galvanometer readings. What kind of current does this generator produce?

Direct current

Name _____ Date _____ Class _____

Chapter 14 Review Worksheet, continued

c. How does this generator differ from the generator shown on page 320 of your textbook?

Be prepared for various responses. This questions requires high-level reasoning. A sample answer is provided on page 324 of the Annotated Teacher's Edition.

d. Explain what is happening in each illustration in the sequence.

In setup (a), the coil is not moving, so no current is produced. In setups (b) and (d), the coil is moving through the magnetic lines of force, causing a current to flow. In setup (c), the parts of the coil in the magnetic lines of force are moving parallel to the lines of force, so no current flows.

2. Current Events

Electricity is related in some way to each of the following energy forms: light, heat, magnetic, chemical, kinetic, and pressure. Identify the relationship among the energy forms in each of the following converters:

- dry cell — **Chemical energy to electrical energy**
- solar cell — **Light energy to electrical and heat energy**
- wet cell — **Chemical energy to electrical energy**
- light bulb — **Electrical energy to heat and light energy**
- generator — **Mechanical and magnetic energy to electrical energy**
- piezoelectric crystals — **Mechanical, or vibrational, energy to electrical energy**
- thermocouple — **Heat energy to electrical energy**

Chapter 14

Name _____ Date _____ Class _____

Chapter 14 Review Worksheet, continued

3. Play It Either Way

A light bulb can use either AC or DC. Hypothesize why this is so.

A light bulb can use either AC or DC current to heat the filament because a resistor is an unpolarized device. It will heat up and glow regardless of the direction of the current flowing through it.

4. Irregular Exercise

Stan made a jump rope out of a loop of wire and then performed the activity pictured. The galvanometer showed that a current was being generated.

Galvanometer

Tape

Wire

Illustration also on page 325 of your textbook

a. Explain what happened. (Hint: What makes a compass work?)
Moving the wire through the magnetic field of the Earth caused a current to flow in the wire.

b. Did Stan generate direct current or alternating current? Explain.
Stan generated alternating current because as the wire moved upward, it generated a current in one direction, and as the wire moved downward, it generated a current in the opposite direction.

Answers • Chapter 14

Name _____ Date _____ Class _____

Chapter 14 Assessment, continued

5. Imagine riding a bicycle after dark. Name one possible problem with using a bicycle dynamo instead of a battery to power the bicycle's light.

Answers will vary. Possible answers include the following: When the bicycle is stopped, there is no current generated, so the light goes out. The dynamo may not produce enough energy for a powerful beam.

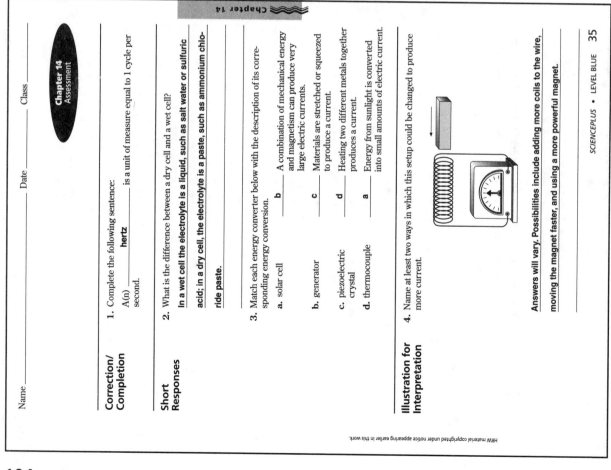

CHALLENGE 1
Short Essay

CHALLENGE 2
Illustration for Interpretation

6. Will this "tree battery" produce continuous electric current? Explain why or why not.

— Galvanometer

— Zinc spike

Copper spike

— Lemon tree

No. The lemons that are supposed to be the electrolytes are not connected to each other, so there is no continuous path for the electrons to follow.

Name _____ Date _____ Class _____

Chapter 14
Assessment

Correction/ Completion

1. Complete the following sentence:

   A(n) _____**hertz**_____ is a unit of measure equal to 1 cycle per second.

Short Responses

2. What is the difference between a dry cell and a wet cell?

   In a wet cell the electrolyte is a liquid, such as salt water or sulfuric acid; in a dry cell, the electrolyte is a paste, such as ammonium chloride paste.

3. Match each energy converter below with the description of its corresponding energy conversion.

   a. solar cell   **b**   A combination of mechanical energy and magnetism can produce very large electric currents.

   b. generator   **c**   Materials are stretched or squeezed to produce a current.

   c. piezoelectric crystal **d**   Heating two different metals together produces a current.

   d. thermocouple   **a**   Energy from sunlight is converted into small amounts of electric current.

Illustration for Interpretation

4. Name at least two ways in which this setup could be changed to produce more current.

   Answers will vary. Possibilities include adding more coils to the wire, moving the magnet faster, and using a more powerful magnet.

≋≋ Chapter 14

Name _____ Date _____ Class _____

EXPLORATION 1

A Conduction Problem to Investigate, page 329

Cooperative Learning Activity

Group size	3 to 4 students
Group goal	to identify factors that affect the conductivity of a wire
Individual responsibility	Each group member should choose a role such as chief investigator, checker, recorder, or materials manager.
Individual accountability	Each group member should be able to complete Applications of Resistance individually.

Safety Alert!

The Problem: Do all wires conduct an electric current equally well? Here are some questions to answer as you investigate this problem.

a. Does the kind of metal affect the transmission of current?
b. What is the effect of having different thicknesses of wire?
c. Does the length of wire influence the current?
d. What happens if a wire resists the flow of current?

You Will Need

- magnet wire
- sandpaper
- thin and thick nichrome wire
- a wooden dowel
- D-cells
- a flashlight bulb
- newspaper
- a coin
- a rubber band
- wire cutters

Part 1: Investigating Questions (a) and (b)

What to Do

1. Prepare equal lengths of the three wires that you will test. Sand the enamel off of the last 2 or 3 cm of each end of the magnet wire.
2. Set up the circuit as shown at right using one of the three wires. Observe the brightness of the light.
3. Do the same for each of the other wires. Observe the bulb in each case. Does the intensity of the light vary? Double-check your results.

Photo also on page 329 of your textbook

Conclusions

1. Is it easier for a current to flow through thin nichrome wire or thin copper wire?

Thin copper wire

Chapter 15

Name _____ Date _____ Class _____

Exploration 1 Worksheet, continued

2. Is it easier for a current to flow through thin nichrome wire or thick nichrome wire?

Thick nichrome wire

3. How might you explain your observations?

It is easier for a current to flow through copper because copper is a better conductor. A thick wire is also a better conductor because there is room for more electrons to pass through—in much the same way that a large hose allows water to pass through more easily than does a thin hose.

Part 2: Investigating Question (c)

What to Do

1. Vary the length of the thin nichrome wire in the circuit by placing the contact wires at different points along the wire, as shown at left.
2. Observe the bulb.

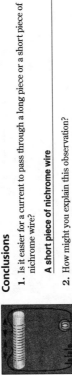

Illustration also on page 329 of your textbook

Conclusions

1. Is it easier for a current to pass through a long piece or a short piece of nichrome wire?

A short piece of nichrome wire

2. How might you explain this observation?

Accept all reasonable responses. Students may infer that certain qualities make some types of wire better conductors. The longer the wire, the greater the amount of resistance.

Interpreting Parts 1 and 2

Some wires do not allow electric charges to move through them as readily as do other wires. In other words, these wires offer more *resistance* to the flow of the charges. Which offers more resistance:

nichrome or copper wire? **Nichrome wire**

thin or thick wire? **Thin wire**

long or short wire? **Long wire**

Name _____ Date _____ Class _____

Exploration 1 Worksheet, continued

Part 3: Investigating Question (d)

What to Do

1. Connect the circuit as shown on page 330 of your textbook.
2. Wrap the wire with newspaper, and then watch it for 1 minute.
3. Now disconnect the circuit, unwrap the newspaper, and carefully touch the wire.

Conclusions

1. When a wire resists the flow of current, what happens?

   The wire becomes warm.

2. What energy change is taking place in the nichrome wire?

   Electrical energy is converted into heat energy.

Any conductor that offers considerable resistance is called a **resistor** and is represented by the symbol ‑w‑. In the space below, use circuit symbols to draw a circuit containing two D-cells, a coil of nichrome wire, a switch, and a bulb.

Sample diagram:

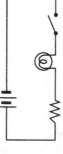

Applications of Resistance

You have found that resistors produce heat, and you have inferred that they reduce the flow of current. Both of these characteristics are useful.

• Resistors are used to produce heat in appliances like toasters or irons. Appliances that produce heat when an electric current flows through them are called *thermoelectric* devices.

Why is this a good name for them?

The prefix *thermo* comes from a Greek word meaning "heat." The term *thermoelectric* means "electricity from heat."

Make a survey of the thermoelectric devices that are used in your home. How much power do they use? (This will be the number, in watts, marked on the appliance.)

Answers will vary. Answers should be reported in watts.

Name _____ Date _____ Class _____

Exploration 1 Worksheet, continued

Resistance can be compared to friction. In what ways do you think they are similar?

They both slow things down, and they both convert one kind of energy into heat energy. Friction converts kinetic energy into heat energy; resistance converts electrical energy into heat energy.

Friction

1. Flick a coin across a table with your finger. It moves a little, slows down, and stops. What causes the coin to slow down?

   Friction between the coin and the surface causes the coin to slow down.

2. Where does the kinetic energy of the moving coin go?

   The coin's kinetic energy is converted into heat energy due to friction.

3. Here's how to find out. Place your finger firmly on the coin and rub it back and forth a dozen times or so against a table top. Now touch the coin to your chin. What kind of energy was produced?

   Heat energy

4. What caused it to be produced?

   Friction

Resistance

Resistance is like friction. Electrons flow because they receive electric energy from a cell. As the electrons flow through a piece of nichrome wire, the wire resists the flow of the electrons—in much the same way that the nails resist the rolling marbles in the photo at left.

1. If the nails were a little closer together, how would this affect the rolling marbles?

   The rolling marbles would slow down.

2. Would this situation represent a wire with more resistance or less?

   More resistance

3. What form of energy do you predict will be produced from the electric (kinetic) energy of the electrons as they slow down?

   Heat energy

You will check your prediction in Part 3 of this Exploration.

Photo also on page 330 of your textbook

Name _____ Date _____ Class _____

Things Are Heating Up

Do this activity after completing Exploration 1 on page 329 of your textbook.

In an experiment to measure the amount of resistance in a circuit, Ian noticed that as the voltage increased, so did the resistance. He also noticed that the temperature of the filament in the bulb increased with the increased voltage. Ian wondered if temperature increase had any effect on the resistance in a circuit. He did some research and found the following data, which show the effect of increased temperature on the resistance of a length of tungsten wire. The filament in the bulb is made of tungsten. Resistance is measured in units called ohms. Plot Ian's data on the grid provided, and then answer the questions on the next page. Don't forget to label the *x*- and *y*-axes.

Temperature (°C)	Resistance (ohms)
27	549
127	783
227	1025
327	1285
427	1562
527	1845
627	2042
727	2421
827	2713
927	3008

Resistance (ohms) vs. Temperature (°C)

Name _____ Date _____ Class _____

Exploration 1 Worksheet, continued

• Resistors also reduce current flow. A *variable resistor*, such as the one used in Part 2 of this Exploration, varies the current. Such resistors are common—the volume control on a radio is one example.

Here is a project for you to try. Turn an ordinary graphite pencil into a usable rheostat (device for varying the current). What might you use it for?

Almost all devices that require variable amounts of electrical current use

rheostats, including the control dial on an electric oven, the volume dial on

a tape deck, and a dimmer switch on a lamp.

A graphite-pencil rheostat

D-cell

Movable contact

Fixed contact

Illustration also on page 330 of your textbook

How does it work?

A graphite-pencil rheostat allows the user to vary the resistance by ad-

justing the movable contact along the surface of the graphite. The longer

the distance between contacts, the greater the resistance.

SCIENCEPLUS • LEVEL BLUE   **41**

Exploration 2 Worksheet, continued

4. A circuit with two bulbs and two switches, P and Q. When P and Q are closed, both bulbs light up. If either switch is opened, neither bulb lights up.

5. A circuit that contains two switches and two bulbs. If both switches are open, neither bulb lights up. If either one of the switches is closed, both bulbs light up.

6. Analyze your findings.

   a. Identify the series and parallel circuits in each of your designs. **Circuit 1 is a series circuit. Circuit 2 is a parallel circuit. (In circuit 3, bulb A is in series with bulbs B and C, which are parallel. In circuit 4, the switches are in series with other bulbs, but the bulbs could be either in series or parallel with each other. In circuit 5, the switches are parallel with each other, but the bulbs could be either in series or parallel.)**

   b. Is there any parallel circuitry in the room where you are now? How could you find out without having to expose any wiring? **Answers will vary. Sample answer: Yes, the lights in the classroom are wired in parallel. You could find out by unscrewing one bulb and seeing whether the others stay lit.**

Sample diagrams for Part 2 can be found on page S210 of the Annotated Teacher's Edition.

Part 3: Current Questions to Investigate

Remember: The brightness of the light bulb is a measure of the amount of current flowing.

You Will Need

- 6 light bulbs
- 3 contact switches
- 3 D-cells
- copper wire
- wire cutters

What to Do

Construct each of the circuits shown in the table on the next two pages, and record the results in the right column. Make certain that switches are included in the circuits you construct.

Putting It Together

On the next page are a number of statements with options. Each statement, with the correct option, is a valid conclusion for one of the four experiments in the table on the next two pages. Choose appropriate conclusions for each experiment and include them in your table.

Graphing Practice Worksheet, continued

1. What does the graph show? Do you think that there is a relationship between resistance and temperature? **The graph shows that as the temperature increases, so does the resistance. There does seem to be a direct relationship between them.**

2. Suppose that the data on the previous page is for a piece of tungsten wire 1 m long. What would be the result of doubling the length of the wire? Why do you think this would happen? **The resistance would double because it would double the number of obstacles to the flow of electrons.**

3. What do you think would happen to the resistance if the temperature of the wire was cooled toward −273°C? (Note: −273°C is the lowest temperature that matter can have. This temperature is also known as absolute zero.) **The resistance of the metal would gradually approach zero also.**

   You would not want zero resistance in a light bulb filament because the filament would not glow. For what uses would zero resistance be beneficial? **One possible response is in wires that transmit electricity. If the resistance in these wires approached zero, then very little energy would be wasted, and energy costs could be greatly reduced. (You may wish to explain to students that materials with zero resistance are called superconductors.)**

Name _____ Date _____ Class _____

Exploration 2 Worksheet, continued

a. Connecting bulbs one after another in a circuit (decreases, increases) the amount of current flowing in the circuit.

b. If the number of cells is increased in the circuit, the amount of current is (decreased, increased).

c. If more bulbs are connected in series in a circuit, the resistance of a circuit is (decreased, increased).

d. If two bulbs are placed in a branched circuit rather than in an unbranched circuit, the current through the battery is (decreased, increased).

e. The current flowing through each bulb in a parallel circuit is (greater than, less than) the current flowing through each bulb in a series circuit.

f. The resistance of a circuit is decreased when the bulbs are placed in (series, parallel).

g. The resistance of a circuit is (more, less) with three bulbs in series than with two bulbs in parallel connected to a third in series.

Question	Experimental design	Results/conclusions of experiment (Select from list above)
1. How is the amount of current affected by the number of cells in a circuit?		**Result:** The bulb with two cells was brighter than the bulb with one cell. The bulb with three cells was brighter than the bulb with two cells. **Conclusion(s): (b)** If the number of cells is increased in the circuit, the amount of current is increased.
2. How is the amount of current affected by the number of bulbs connected in series, that is, one right after the other?		**Result:** When two bulbs were connected in series, they were dimmer than when there was only one bulb. When three bulbs were connected in series, they were dimmer than when there were only two bulbs. **Conclusion(s): (a)** Connecting bulbs one after another in a circuit decreases the amount of current flowing in the circuit. **(c)** If more bulbs are connected in series in a circuit, the resistance of a circuit is increased.

48   UNIT 5 • ELECTROMAGNETIC SYSTEMS

Name _____ Date _____ Class _____

Exploration 2 Worksheet, continued

Question	Experimental design	Results/conclusions of experiment (Select from list on the previous page)
3. How does the current flowing through each bulb in a *parallel*, or branched, circuit compare with the current flowing through each bulb in a series circuit?		**Result:** The two bulbs in the parallel circuit were brighter than the two bulbs in the series circuit. **Conclusion(s): (e)** The current flowing through each bulb in a parallel circuit is greater than the current flowing through each bulb in a series circuit. **(f)** The resistance of a circuit is decreased when the bulbs are placed in parallel.
4. What difference is there between the current flowing through the battery when two bulbs are connected in series and the current flowing through the battery when two bulbs are connected in parallel?		**Result:** All of the bulbs were dimmer when they were connected in series than when they were connected in parallel. **Conclusion(s): (d)** If two bulbs are placed in a branched circuit rather than in an unbranched circuit, the current through the battery is increased. **(g)** The resistance of a circuit is more with three bulbs in series than with two bulbs in parallel connected to a third in series.

SCIENCEPLUS • LEVEL BLUE   49

Chapter 15

Answers • Chapter 15

Circuit Mastery

Try this activity after you complete Lesson 3, Controlling the Current, which begins on page 335 of your textbook.

If you can complete these drawings, you will be current on circuits.

a. Modify the circuit by adding what is needed to light the bulb.

b. Modify the circuit by adding what is necessary to keep the cell from being quickly drained of its electrical energy.

c. Modify the circuit by adding what is necessary to turn each light on and off independently.

d. Modify the circuit by adding what is necessary to turn the light on with switches located in two different places.

e. Modify the circuit by adding something to the circuit to control the brightness of the light.

f. Modify the circuit by adding a switch that will turn all of the bulbs on and off. Then add switches so that bulbs B and C can be controlled individually while bulb A is on.

g. Draw two switches to operate the circuit so that if the bulb is lit when one switch is closed, the other switch will be able to open the circuit—and vice versa. Add any wiring that may be needed.

EXPLORATION 4

Constructing an Electromagnet, page 338

Your goal	to construct an electromagnet strong enough to hold up a number of washers	Safety Alert!

You Will Need

- a paper clip
- washers
- some light, insulated wire
- 2 D-cells
- an iron spike
- a switch

Spike

Paper clip jumps to the spike.

Switch closed

Illustration also on page 338 of your textbook

What to Do

Get together with one or two classmates. Your task will be to make a functioning electromagnet and then to determine how its strength can be increased.

Making the Electromagnet

1. Use the materials illustrated to make an electromagnet capable of supporting a paper clip from which several washers are hanging.

2. Sketch your electromagnet. Then label and trace the path of the current.

   The parts of the circuit include the cell and the insulated copper wire. The current travels from the cell through the wire and back to the cell again.

3. Are the spike, paper clip, and washers part of the circuit?   **No**

4. How many washers can your first design hold?   **Answers will vary.**

Increasing the Electromagnet's Strength

How can you make your electromagnet hold more washers? Take some time to discuss the following:

- factors or variables that might be altered
- ways to measure the magnet's strength
- safety precautions
- the apparatus

Get your design approved by your teacher. Then assemble the necessary apparatus, do the experiment, record the results, and draw conclusions based on your results. Record your results in your ScienceLog and share your results with others.

Allow students to brainstorm their ideas for increasing the strength of the electromagnets. Inference and experimentation should lead them to the conclusion that increasing the strength of the current or increasing the number of turns in the coil will increase the strength of the electromagnet. One simple way to test their ideas is to see how many washers the nail will support after each change is made in the circuit.

Name _____ Date _____ Class _____

Using Electrical Units

Do this activity with Lesson 5, How Much Electricity? on page 343 of your textbook.

Complete the table below.

What we are measuring	Unit	Definition	Problems
quantity of charge	1 coulomb	6.24 quintillion electrons	How many electrons are in 5 C of charge? __31.2 quintillion__
current	1 ampere	1 coulomb of charge per second	a. For 5 A, how many coulombs of charge pass in 2 s? __10 C__ What is the current? __5 A__ b. If 20 C of charge pass in 2 s, what is the current? __10 A__ c. If 4 A of current flow for 5 s, what is the quantity of charge? __20 C__
energy per quantity of charge	1 volt	1 joule of energy per coulomb of charge	a. How much energy per coulomb is given by a 6 V battery? __6 J__ b. How much energy is given to 5 C of charge by a 6 V battery? __30 J__ c. If 6 J of energy are given to 3 C of charge, what is the strength of the battery? __2 V__
power (energy per second)	1 watt	1 joule of electrical energy used in 1 second	a. How many joules of energy are used in 1 s by a 600 W iron? __600 J__ b. How many joules of energy does a 2000 W heater use in 1 min.? __120,000__ c. If a radio uses 1000 J in 10 s, what is its power? __100 W__
power (energy per second)	1 kilowatt	1000 watts	a. 200 W equals how many kilowatts? __0.2 kW__ b. 500 kW equals how many watts? __500,000 W__

Name _____ Date _____ Class _____

Using the Theme of Systems

This worksheet is an extension of the theme strategy outlined on page 348 of the Annotated Teacher's Edition. It is also designed as an extension of Chapter 15, Currents and Circuits, which begins on page 326 of the Pupil's Edition.

Focus question	How is current through a circuit similar to blood flow through the human body?

The human circulatory system is responsible for transporting blood throughout the body. Do some research to learn about the circulatory system, and then answer the following questions:

1. The heart acts as a pump, pushing blood through the arteries at high pressure. What part of a circuit acts as an electron pump?

   The chemical cell

2. Is the flow of blood through the body a "direct current" or an "alternating current"? Explain your reasoning.

   Direct current. Blood can only travel in one direction through the body; it cannot reverse its direction.

3. Would you define the circulatory system as a parallel circuit or a series circuit? Explain your reasoning.

   A parallel circuit. Blood flows through many different paths, such as tissues and organs, at the same time.

4. Current meets with resistance as it flows through a circuit. Where might blood flow experience resistance in the circulatory system?

   Answers will vary. Possible answers include the smaller vessels and capillaries, blood clots, and arteries narrowed by coronary disease.

5. Valves in the heart and veins are similar to diodes, electrical devices that allow current flow in only one direction. Explain this similarity.

   Sample answer: These valves control the flow of blood through the heart and veins. When these valves are closed, blood cannot flow backward. Thus, these "diodes" keep the "current" flowing.

Challenge Your Thinking, page 348

1. Safe Circuit

Many apartment buildings have security doors that can be opened from each apartment. The diagram on page 348 of your textbook shows the circuitry of one such security door. Explain how it works.

Sample answer: When the button is pressed, current flows through the

circuit. The current flowing through the wire causes the bar inside the wire

coil to become magnetized. The electromagnet pulls down the arm at-

tached to a pivot. This removes the lock from the door.

_____

_____

2. Go With the Flow

The drawing on the next page from a sixth-grade science book illustrates a lesson about electricity. Write a paragraph to go with this lesson and then fill in the blanks in the diagram on the next page.

Sample answer: The battery supplies energy to the electrons on the nega-

tive electrode. As the current of electrons moves through the circuit, it

feels resistance first from the wire and then from the resistor, and some of

its electrical energy is changed into heat energy. As the electrons flow

through the highly resistant filament of the lamp, more of their energy is

lost in the form of heat and light energy. Most of the remaining energy in

the current is changed into the mechanical energy of the motor. Depleted

of their energy, the electrons return to the battery, where they are re-

charged with more energy to begin the cycle again.

An Electrifying Magic Square

Try this activity after completing Chapter 15, which begins on page 326 of your textbook.

Select the number of the term in list II that best matches each statement in list I. Place the number in the appropriate space in the magic square. Add the horizontal rows, the vertical columns, and the diagonals to discover the magic number. Not all of the items in list II will be used.

List I

A. A flow of electrical charges
B. One coulomb of charge flowing past a point in a circuit in one second
C. A measure of the force that pushes current through a conductor
D. 1 J/s
E. Actual charged particles flowing in most circuits
F. Unit for electrical charge
G. One joule of electrical energy per coulomb
H. Electrical energy (large quantity)
I. Electrical energy (small quantity)

List II

1. proton
2. ampere
3. joule
4. watt
5. volt
6. electrons
7. voltage
8. coulomb
9. current
10. kilowatt-hour
11. neutrons

A	B	C
9	2	7
D	**E**	**F**
4	6	8
G	**H**	**I**
5	10	3

Magic number = ___18___

Name ______________ Date ________ Class ________

Chapter 15 Review Worksheet, continued

What words should be used to fill in the blanks?

Sample labels:

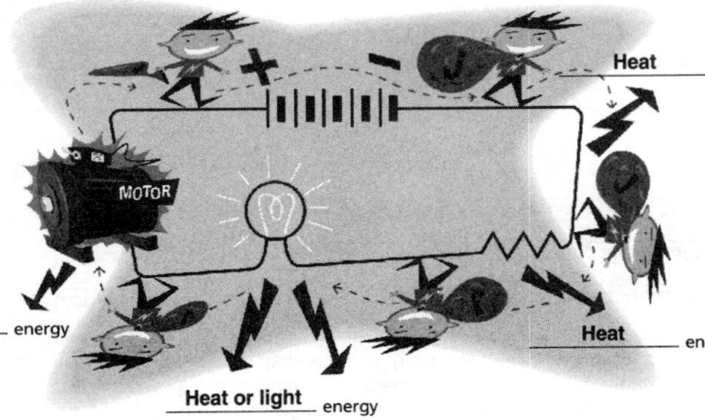
MOTOR
Heat energy
Mechanical energy
Heat energy
Heat or light energy

Illustration also on page 348 of your textbook

3. The Inside Story

A common type of switch is the *dimmer switch*. Turn the knob clockwise, and the light becomes brighter. Turn the knob counterclockwise, and the light becomes dimmer. Draw a sketch showing the circuit that may be inside.

A possible arrangement is as follows:

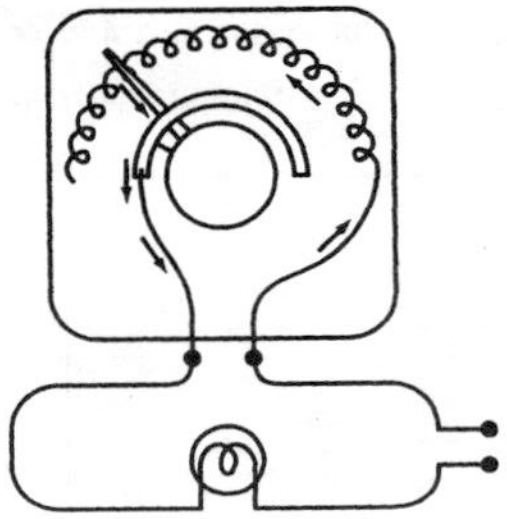
Source of electricity

Name ______________ Date ________ Class ________

Chapter 15 Review Worksheet, continued

4. You Be the Teacher

Ms. Alvarado asked her class to design a circuit containing two switches (S_1 and S_2), two light bulbs (L_1 and L_2), and one dry cell. She asked them to arrange the circuit so that it does the following:

a. If only S_1 is closed, L_1 will light.

b. If only S_2 is closed, nothing will happen.

c. If S_1 and S_2 are closed, both lamps will light.

Their work is shown below.

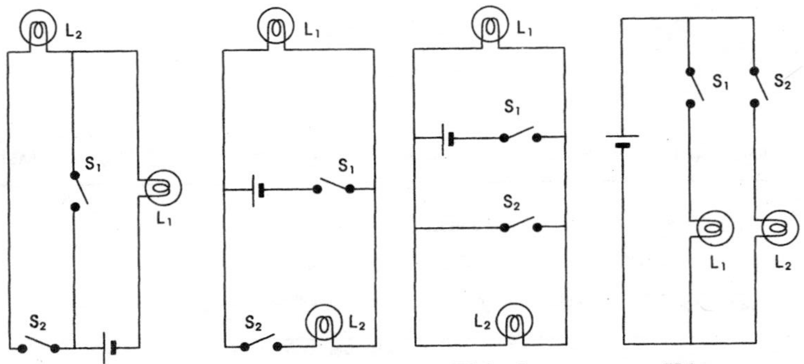
L2
S1
L1
S2
Ruth
L1
S1
S2
L2
Andy
L1
S1
S2
L2
Michael
S1
S2
L1
L2
Yelena

Play the role of Ms. Alvarado, and grade the students' work. Indicate how you know whether each design is right or wrong.

Only Andy's circuit met all three requirements. In Ruth's circuit, only (a) is met. Her circuit fails requirement (b) because L_1 and L_2 would light, and it fails requirement (c) because only L_1 would light. In Michael's circuit, requirement (b) is met but his circuit fails to meet requirement (a) because both L_1 and L_2 would light, and it fails requirement (c) because neither L_1 nor L_2 would light. Requirements (a) and (c) are met in Yelena's circuit, but her circuit fails requirement (b) because L_2 will light.

Chapter 15

Name ______ Date ______ Class ______

Chapter 15 Review Worksheet, continued

5. The Ol' Double Switcheroo

In the circuit shown below, the upper switch moves from contact *A* to contact *C* and the lower switch moves from contact *B* to contact *D*.

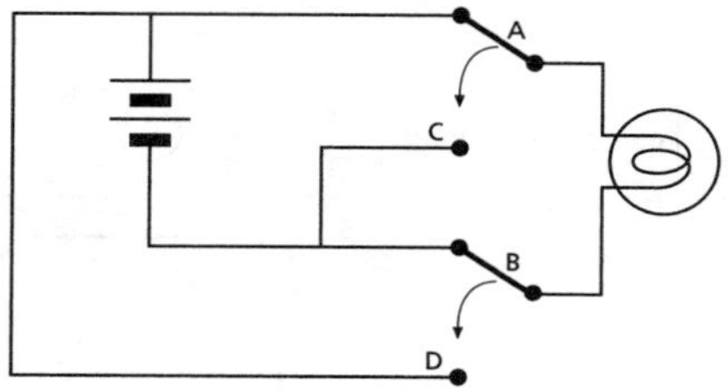

Suggest a function for this circuit.

Answers may vary. One use for this circuit would be to allow two different switches to control one light.

6. Safe at Home

Fuses are designed to protect household circuits against damage due to electrical overload. Fuses are made of metal alloys that have low melting points. When a circuit carries more current than is safe, the fuse strip melts. How do you think fuses work? (Hint: Think about the effects of resistance.)

Sample answer: Fuses are made of materials that offer high resistance to the current passing through them. Therefore, some of the electrical energy is converted into heat energy. When this buildup of heat energy becomes too great, the fuse melts, breaking the circuit.

Name ______ Date ______ Class ______

Chapter 15
Assessment

Word Usage

1. Use all of the following terms in one or two sentences to show how they are related: *amperes, current, joules, coulombs, energy,* and *volts.*

Sample answer: The rate at which electricity flows is called *current*, which is measured in *amperes*, or *coulombs* per second; the *energy* given to a unit of charge flowing in a circuit is measured in *volts*, or *joules* per coulomb.

Correction/ Completion

2. The following sentences are incorrect or incomplete. Your challenge is to make them correct and complete.

   a. In a parallel circuit, if one appliance burns out, all of the other appliances stop working.

   In a *series* circuit, if one appliance burns out, all of the other appliances stop working.

   b. Appliances such as toasters and irons convert **electrical** energy into **heat** energy because their circuits offer a lot of **resistance**.

Answering by Illustration

3. Using the correct symbols, draw a circuit diagram to match this illustration.

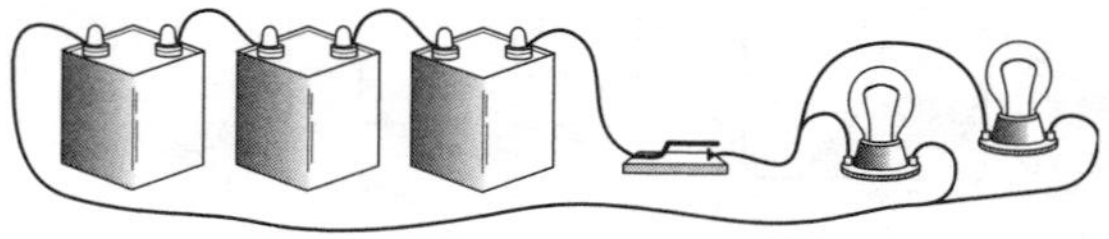

Sample answer:

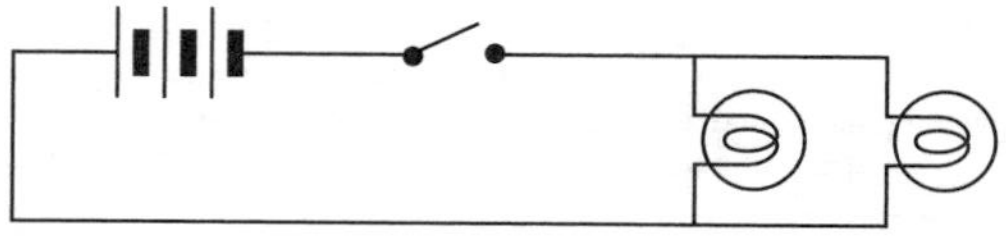

Name _____ Date _____ Class _____

Electrifying News

Lana Lectron used to be an electrical engineer before she became a reporter, which could explain why she sometimes "electrified" some key words and phrases in the story that follows. See if you can spot the electrical terms, and then use them to fill in the blanks in the glossary that follows.

Dangerous Nic Rome was captured by Detectives Ronald and Rhoda Franklin. They work for the A.C.D.C. (Apprehend Criminals Detective Company). Nic Rome has been charged with the conduction of an attractive 18-year-old coil, Millie Volt, and with the theft of valuable joules.

This is watt appears to have happened. The criminal avoided a battery of policemen by escaping on his kinetic cycle. Spotting him from a parallel road, the detectives proton their hats and jumped into their motor car. In an instant, the thermocouple were in hot pursuit. They forced Nic Rome off the road, causing him to crash into a coulomb of poles. Nic grabbed his weapon and swung it with vicious energy to ampere his arrest. The detectives warded off this series resistance and subdued the potential killer.

Nic Rome is now in solartary confinement in the local chemical cell. His trial will be held in the current court session presided over by the circuit judge. The community is relieved at the capture of the crime generator. As a result, they have agreed to elect Ron and Rhoda Detectives of the Year.

Glossary of Electrical and Magnetic Words

a. Units

1. Unit of electric charge _____ Coulomb

2. Unit of current; one coulomb per second _____ Ampere

3. Unit of energy _____ Joule

4. One joule per coulomb _____ Volt

5. One joule of energy used in one second _____ Watt

b. Energy and Energy Converters

6. Type of energy possessed by a raised object _____ Potential

7. Energy of moving objects _____ Kinetic

8. Device that converts electrical energy into mechanical energy _____ Motor

9. Device in which electrical energy is produced by chemical action _____ Chemical cell

10. Device in which a current is produced by a magnet revolving near a coil of wire _____ Generator

11. Group of cells _____ Battery

12. Sensitive device for measuring temperature _____ Thermocouple

---

Name _____ Date _____ Class _____

Chapter 15 Assessment, continued

Numerical Problem

4. A 400 W hair dryer is connected to a 110 V household current.

a. How many joules of energy are applied to each coulomb of charge?

110 V = 110 J/C, so 110 J of energy are applied to each coulomb of charge.

b. If the hair dryer is used for 3 minutes, how many joules of energy are used?

400 W = 400 J/s × 60 s/min. × 3 min. = 72,000 J of energy used in 3 min.

c. How many coulombs were required during those 3 minutes?

72,000 J ÷ 110 J/C, or about 655 coulombs, were required.

d. If the charge was constant during the 3 minutes, what was the rate of electrical flow per second?

655 C/3 min. × 1 min./60 sec. = about 3.6 C/s, or 3.6 A, of current

CHALLENGE 1

Answering by Illustration

5. Draw an illustration to match this circuit diagram.

Sample answer:

CHALLENGE 2

Short Essay

6. Explain why modern homes use parallel circuits rather than series circuits.

Sample answer: Modern homes use parallel circuits because they offer two main advantages over series circuits: current is not reduced as more appliances are added to the circuit, and the current is not broken if one appliance burns out.

≋ **Answers • Unit 5**

Electrifying News, continued

c. Circuits

13. The easy passage of electric charges through a material	Conduction
14. Path along which electrons flow	Circuit
15. Type of friction met by electric charges moving in a conductor	Resistance
16. Type of circuit in which appliances are connected one after the other	Series
17. Type of circuit with branches	Parallel
18. Part of an electromagnet	Coil
19. A current that moves in one direction and then in the opposite direction	AC
20. A current that moves only in one direction	DC
21. Rate of flow of electric charges in a conductor	Current
22. A type of wire with high resistance to the flow of current	Nichrome
23. One reversal of current	Cycle

d. Charges and Cells

24. Smallest particle with negative charge	Electron
25. Small particle with positive charge	Proton
26. Conducting terminals in a cell	Electrodes
27. Given to electrons by a chemical cell	Energy
28. Type of cell used to obtain electrical energy from light energy	Solar

e. Magnets

29. Type of force between the north and south poles of a magnet	Attractive
30. Parts of a magnet where the strength is greatest	Poles

Making Connections, page 350

The Big Ideas

In your ScienceLog, write a summary of this unit, using the following questions as a guide:

1. How may electricity be produced? (Ch. 13)
2. What is the difference between a cell and a battery? (Ch. 13)
3. What parts are common to all chemical cells, and how do they operate? (Ch. 13)
4. How do AC and DC currents differ? (Ch. 14)
5. What factors affect the size of currents in circuits? (Ch. 15)
6. What are some differences between series and parallel circuits? (Ch. 15)
7. What are electromagnets? How can their strength be increased? (Ch. 15)
8. How is the flow of water like that of electricity? (Ch. 15)
9. By what units is electricity measured? How are these units related? (Ch. 15)

A sample unit summary is provided on page 350 of the Annotated Teacher's Edition.

Checking Your Understanding

1. Recall your introduction to fuses in Challenge Your Thinking, Chapter 15, on page 349 of your textbook. Suppose a fuse is rated at 15 amps (it melts when the circuit carries more than 15 amps). How many 100 W light bulbs would it take to blow the fuse? Assume a 110 V current. (Don't try this yourself!) You will find questions 6 and 7 on page 346 of your textbook to be helpful.

Watts are equal to volts times amperes. 110 V × 15 A = 1650 W

1650 W ÷ 100 W/light bulb = 16.5 light bulbs

Therefore, 17 light bulbs would be needed.

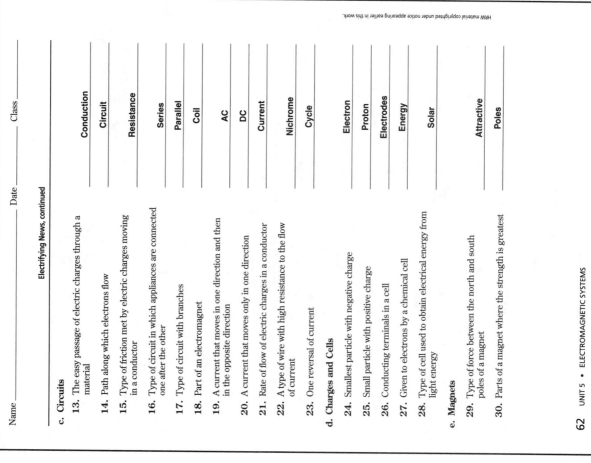

Top view of
house fuse

Side view of
house fuse

Fuse
strip

Illustration also on page 350
of your textbook

Name _____ Date _____ Class _____

Unit 5 Review Worksheet, continued

4. The diagrams below show two different types of microphones.

Diaphragm
Wire coil
Electric current
Magnets
Sound waves

Diaphragm
Electric current
Piezoelectric crystal
Sound waves

Illustration also on page 351 of your textbook

Use the diagrams and your knowledge of the principles of electricity to explain how each of these microphones works.

In the microphone on the left, the vibrating diaphragm causes a wire coil to move across a magnetic field. As the wire cuts through the field, a current is produced in the wire. In the microphone on the right, sound waves hit the diaphragm, causing it to vibrate. This vibration squeezes the crystals, which produces an electric current. This current corresponds to the original sound.

5. Make a concept map using the following terms: volts, current, electricity, amperes, voltage, coulombs, and charge.

Sample concept map:

Electricity
consists of
current — measured in — amperes
voltage — measured in — volts
charge — measured in — coulombs

Name _____ Date _____ Class _____

Unit 5 Review Worksheet, continued

2. Below is a diagram of a circuit breaker, a device that mechanically performs the same task as a fuse, breaking the circuit when too much current flows through it. How does this device work?

Trip lever
Closed contacts
Current path
Bimetallic strip
Fixed position in the circuit-breaker box

2. Contacts open
3. Current flow stops
1. Strip bends

A. Normal current   **B. Overload current**

Illustration also on page 351 of your textbook

Sample answer: As the current flows through the circuit, the bimetallic strip offers some resistance to the current flow, causing the strip to heat up due to friction. Because the two heated metals expand at different rates, the strip bends. If the current is great enough, the strip bends enough so that it no longer holds the contacts together, and the circuit is broken.

3. Electricity is often compared to flowing water. Read the following examples and decide whether each suggests high or low amperage, high or low voltage, or any combination of these.

a. the Mississippi River

High current, low voltage

b. Niagara Falls

High current, high voltage

c. the blast from the spray-gun nozzle at a do-it-yourself car wash

Low current, high voltage

d. a dripping faucet

Low current, low voltage

Unit 5
End-of-Unit Assessment

Word Usage

1. Write a sentence about each of the situations described below. In each sentence, use at least one of these terms: *resistance, dry cells, wet cells, series, kilowatt-hours,* and *circuit.*

   a. The heating element in the toaster glows red.

   Sample answer: The resistance of the wire in the toaster causes the heating element to heat up.

   b. The meter reader checked the meter on our house today.

   Sample answer: The meter reading showed the amount of electrical energy consumed in kilowatt-hours.

Correction/Completion

2. Correct the statements below.

   a. A battery consists of a single cell.

   Sample answer: A battery consists of a number of cells connected together.

   b. Energy flow is always counterclockwise in an alternating current generator.

   Sample answer: Energy flow changes direction in an alternating current generator.

   c. A series circuit has at least two branches, with the current divided between each.

   Sample answer: A parallel circuit has at least two branches, with the current divided between each.

Short Responses

3. Of the following items, circle the ones that would make good conductors of electricity.

   plastic ruler   (safety pin)   plastic comb

   wooden pencil   (scissors)   (nail)

   plastic wrap

Unit 5 Assessment, continued

4. Match each of the following terms with the proper definition or description.

   a. ampere   __c__   a unit of one cycle per second

   b. power   __d__   a unit of measure of electrical charge

   c. hertz   __a__   1 C passing a point in 1 s

   d. coulomb   __e__   the energy of the charge flowing in a circuit

   e. voltage   __b__   the rate at which electrical energy is used

   f. current   __f__   flow of charge in a conductor

5. Something is missing from each of the drawings. Sketch and label the missing parts.

   a. electromagnet

Illustration for Correction or Completion

   There needs to be a moving magnet inside the coil.

   b. complete circuit

   An energy source, such as a cell or a collection of cells, needs to be added. Also, the switch must be closed in order for an electrical current to flow through the circuit.

Name ______________ Date ________ Class ________

Unit 5 Assessment, continued

Illustration for Interpretation

6. Use the illustrations below to answer the questions that follow.

Circuit 1

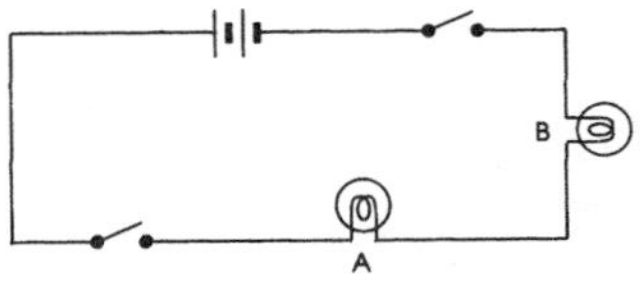

Circuit 2

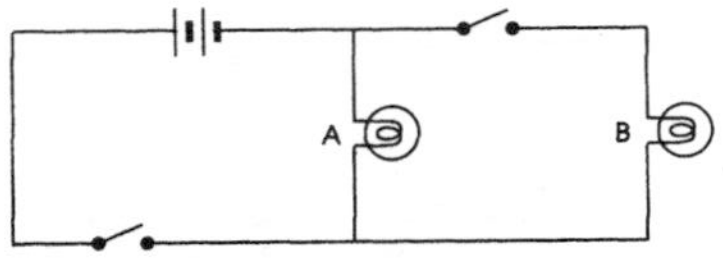

a. How is Circuit 1 different from Circuit 2?

Circuit 1 is a series circuit. Circuit 2 is a parallel circuit.

b. If you want to be able to use light *A* without light *B*, which circuit would you use?

Circuit 2

c. In Circuit 1, does either bulb glow more brightly than the other? If so, which one?

The bulbs glow equally bright.

d. In Circuit 2, does either bulb glow more brightly than the other? If so, which one?

The bulbs glow equally bright.

CHALLENGE 1

Numerical Problem

7. A toaster, electric kettle, and coffee maker are all being operated from the same electrical outlet.

Information:

kettle	1500 W	
toaster	800 W	
coffee maker	450 W	

The house operates on a 110 V circuit.

Would a 20 A fuse blow in this case? Show your work.

Yes. 1500 W + 800 W + 450 W = 2750 W

2750 J/s ÷ 110 J/C = 25 C/s, or 25 A, required

Name ______________ Date ________ Class ________

Unit 5 Assessment, continued

Graph for Interpretation

8. Use the graph below to answer the questions that follow.

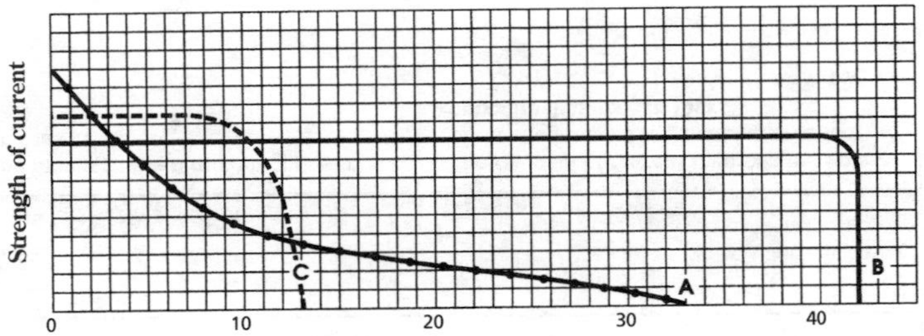

A = alkaline cell B = mercury cell C = ordinary cell

a. Which cell has the shortest life?

***C* (ordinary cell)**

b. How many hours will the mercury cell last?

About 42 hours

c. Which is the strongest cell when new?

***A* (alkaline cell)**

CHALLENGE 2

Answering by Illustration

9. Using the principles of an electromagnet, design and draw a model crane that you could use to pick up thumbtacks from the floor. Clearly label your illustration to show how it works.

Equipment: cells (2) or battery, wires, pulley and cable, crank, nail, switch

Sample illustration:

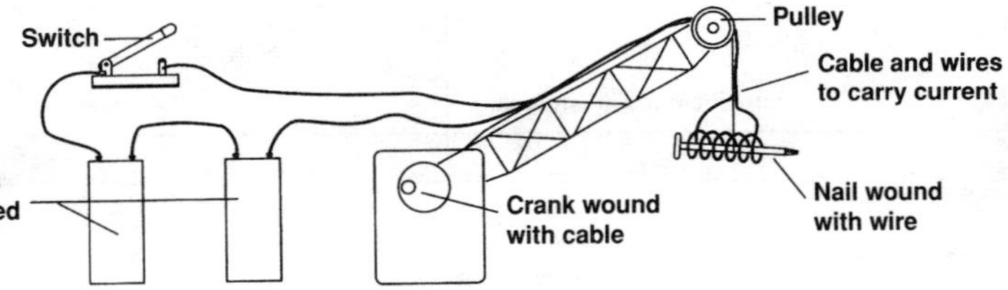

Name _____ Date _____ Class _____

Activity Assessment, continued

Data Chart

Task 1

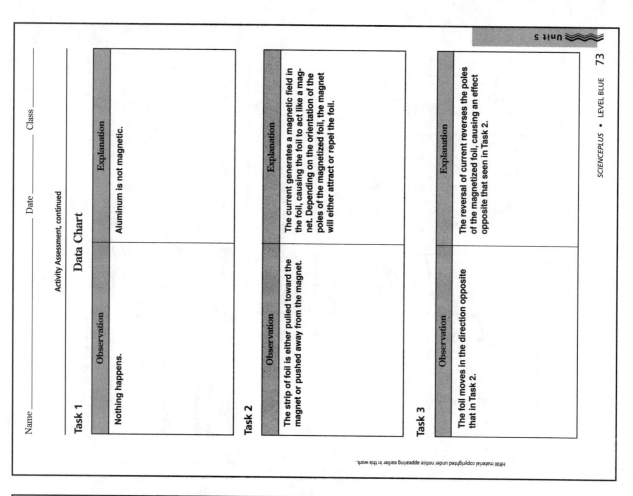

Observation	Explanation
Nothing happens.	Aluminum is not magnetic.

Task 2

Observation	Explanation
The strip of foil is either pulled toward the magnet or pushed away from the magnet.	The current generates a magnetic field in the foil, causing the foil to act like a magnet. Depending on the orientation of the poles of the magnetized foil, the magnet will either attract or repel the foil.

Task 3

Observation	Explanation
The foil moves in the direction opposite that in Task 2.	The reversal of current reverses the poles of the magnetized foil, causing an effect opposite that seen in Task 2.

Name _____ Date _____ Class _____

Unit 5 Assessment, continued

Illustration for Correction or Completion

10. If necessary, correct each of the following circuits by adding, removing, or changing a wire to make a bulb light. If the circuit is already complete, write a C inside it.

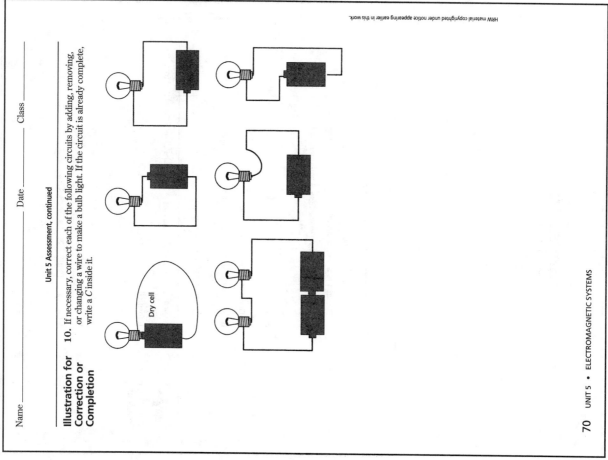

Name ______________ Date ________ Class ________

Unit 5 SourceBook Activity Worksheet

Making a Compass

Complete this activity after reading pages S91–S96 of the SourceBook.

You Will Need

- a large sewing needle
- a piece of cork
- a magnet
- tape
- a small dish
- a compass
- water

What to Do

1. Hold the needle carefully by the blunt end. Hold the magnet with your other hand and move one pole along the length of the needle in the same direction at least 25 times. Explain the effect this action has on the needle.

   It magnetizes the needle by aligning the domains in the metal.

2. Construct a compass by taping the needle horizontally to the piece of cork. Then float the cork in a small dish of water. Try pointing the needle in different directions and releasing it. What happens? Explain.

   The needle always points in the same direction. Because the Earth has a magnetic field, a compass always lines up with the Earth's magnetic field, pointing north and south.

3. Use the commercial compass to check the results obtained with your homemade compass. Were your results accurate?

   Responses will vary. The commercial compass should verify the results.

4. Name two factors that might cause the results to be different.

   Possible answers include the following: The needle was not fully magnetized; the magnet used to magnetize the needle had lost some of its magnetic power; the domains lost their alignment because of some other magnetic influence.

5. Name the three magnets involved in this activity, and describe their alignment.

   One magnet is the needle in the homemade compass. A second magnet is the needle of the commercial compass. The third magnet is the Earth, which behaves as if a giant bar magnet were running through its center. The magnetic fields of all of these magnets align north to south.

Name ______________ Date ________ Class ________

Unit 5 SourceBook Review Worksheet

Unit CheckUp, page S103

Concept Mapping

The concept map shown here illustrates major ideas in this unit. Complete the map by supplying the missing terms. Then extend your map by answering the additional question below.

Sample concept map:

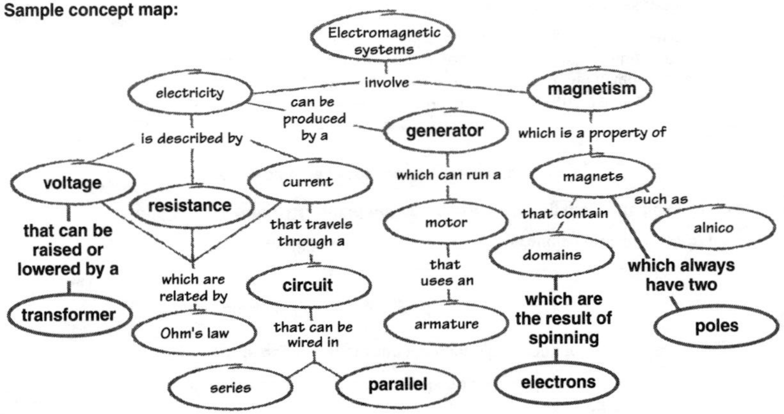

Where and how would you connect the terms *electrons*, *poles*, and *transformer*?

Checking Your Understanding

Select the choice that most completely and correctly answers each of the following questions.

1. An electric current results from
   - a. the flow of positively charged particles through a conductor.
   - b. (the flow of electrons between areas of greater and lesser charge.)
   - c. the movement of ions through wires.
   - d. the flipping of a switch.

Name _____ Date _____ Class _____

SourceBook Review Worksheet, continued

2. A certain transformer has a primary coil with 30 turns of wire and a secondary coil with 10 turns; the voltage in the secondary coil is 30 V. What is the voltage in the primary coil? Show your work.

$E_P/E_S = T_P/T_S$, so $E_P = 30 V \times 30 \text{ turns} \div 10 \text{ turns} = 90 V$

3. Suppose you coupled a generator to an electric motor so that the motor drove the generator and the generator supplied the motor with electricity. Would this system run itself indefinitely? Why or why not?

No; friction and electrical resistance, no matter how small, will eventually remove enough energy from the system to cause it to stop.

4. Why is AC, but not DC, easily transformed?

AC is easily transformed because it is easy to change its voltage.

What would happen if you passed DC through a transformer? Explain.

Nothing. DC would not produce a changing magnetic field, so no current would be produced in the secondary coil.

5. All magnets have a *Curie point*, the temperature at which the materials cease to be magnetic. Above that temperature, no magnetism is present. As you know, there is a relationship between temperature and molecular behavior. How does this relationship help to explain the Curie point?

As the temperature of a material rises, its atoms move more vigorously. At the Curie point, the arrangement of the atoms and molecules in a material are affected by this motion in such a way that domains cannot be formed. Without domains, in which atoms are aligned magnetically, there can be no magnetism in the material.

Name _____ Date _____ Class _____

SourceBook

SourceBook Review Worksheet, continued

2. Which of the following correctly describes the behavior of charged particles?
   a. Negatively charged particles attract positively charged particles.
   b. Negatively charged particles attract other negatively charged particles.
   c. Positively charged particles attract other positively charged particles.
   d. Negatively charged particles attract both positively charged and negatively charged particles.

3. Voltage can be compared to
   a. flow rate.
   b. current velocity.
   c. pressure.
   d. friction.

4. "The greater the distance between two objects, the weaker their interaction," is a statement that applies to
   a. only an electric force.
   b. only a magnetic force.
   c. both an electric and a magnetic force.
   d. neither an electric nor a magnetic force.

5. A generator converts
   a. electrical potential into electrons.
   b. electrical energy into mechanical energy.
   c. direct current into alternating current.
   d. mechanical energy into electrical energy.

Critical Thinking

Carefully consider the following questions, and write a response that indicates your understanding of science.

1. Suppose that you have an iron nail and a compass. How could you demonstrate whether the nail was magnetized or not?

Sample answer: Both poles of the compass would be attracted to the end of an unmagnetized iron nail. If the nail were magnetized, it would repel one of the poles of the compass.

SCIENCEPLUS • LEVEL BLUE 77

Name _____ Date _____ Class _____

Unit 5 SourceBook
Assessment

1. If a neutral object loses electrons, it will
   a. (become positively charged.)   b. become negatively charged.
   c. remain neutral.

2. Which of the following would result in a static charge?
   a. like charges attracting one another
   b. like charges repelling one another
   c. (electrons collecting on the surface of an object)
   d. electrons moving through a conductor

3. Only the distance between two charged objects affects the strength of an electric force.
   a. true   b. (false)

4. According to Coulomb's law, if the distance between two charged objects is decreased, the electric force will
   a. remain the same.   b. decrease.   c. (increase.)

5. Voltage is a measure of
   a. current.   b. (potential difference.)   c. wattage.   d. magnetism.

6. The filament in a light bulb gives off heat in addition to light because of the _____ in the filament.
   a. potential difference   b. pressure   c. voltage   d. (resistance)

7. Ohm's law states that the _____ increases with increasing _____ but decreases with increasing resistance.
   a. voltage, power   b. static electricity, amperage
   c. (electric current, voltage)   d. potential difference, electric current

8. Magnetism is not evenly distributed in a magnet.
   a. (true)   b. false

9. When you cut a bar magnet in half, one half acts as the south pole of the magnet and one half acts as the north pole of the magnet.
   a. true   b. (false)

10. Magnetic lines of force
   a. often cross each other.
   b. are straight lines at the end of a magnet.
   c. have no effect on a compass needle.
   d. (can be used to predict which way a compass needle will point.)

80   UNIT 5 • ELECTROMAGNETIC SYSTEMS

Name _____ Date _____ Class _____

Interpreting Photos

SourceBook Review Worksheet, continued

Some of the bulbs in this string of lights are not working. What type of circuit—series or parallel—would account for the on-and-off pattern of bulbs shown here? Explain.

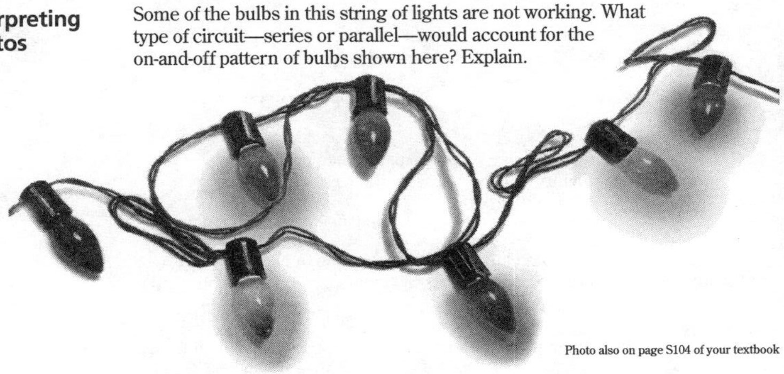

Photo also on page S104 of your textbook

A parallel circuit would account for the on-and-off pattern of bulbs. Because two bulbs are not working, one or both must be faulty. If the wiring were composed of a series circuit, one faulty bulb would cause all of the other bulbs to go out. Two series circuits would show every second bulb out due to a single faulty bulb, but that pattern is not indicated. A burned-out bulb in a parallel circuit does not affect any other light in the string. If the string of lights is wired in parallel, both unlit bulbs must be faulty.

_____

_____

_____

Portfolio Idea

Imagine that you are a scientist in prehistoric times, and you have just discovered magnetism. You are describing its properties to a colleague in a letter. Since you have discovered it well before the modern age of science, most of the words associated with magnetism do not yet exist. In your ScienceLog, describe your discoveries about magnets and magnetism to your colleague without using the terms *magnet, magnetism, pole, magnetic field,* or *lines of force.*

Students' letters should be imaginative as well as scientifically accurate. Check to make sure that they understand magnetism enough to describe it with alternative terms.

SCIENCEPLUS • LEVEL BLUE   79

Name _____ Date _____ Class _____

SourceBook Assessment, continued

20. How could hitting a temporary magnet cause it to lose its magnetic properties?
It could cause the magnetic domains to become randomly arranged, which would cause them to cancel each other's magnetic properties.

21. Explain how a spark can be released when you walk on a rug and then touch a doorknob.
When you walk on a rug, your feet may pick up electrons from the atoms of the carpet. This gives your body a negative electric charge. When you get near an uncharged object, like the doorknob, extra electrons flow from you to the object and cause a small spark.

22. Distinguish between a step-down transformer and a step-up transformer in terms of voltage.
A step-down transformer reduces the voltage across a circuit, and a step-up transformer increases the voltage across a circuit.

23. List two of the four things you must have in order to use electricity.
Answers should include two of the following: a source of electricity, a way to manipulate voltage, a device for converting electricity into useful work, and a pathway for the current.

24. How much current (in amps) is needed to light a bulb that has 0.5 Ω of resistance using a 1.5 V battery? (Show your work.)
$I = E/R$; $I = 1.5$ V$/0.5$ Ω $= 3$ A

25. What is the voltage of the current leaving the transformer in the situation described below? Show your work.
turns in the primary coil (T_P) = 120
turns in the secondary coil (T_S) = 480
voltage in the primary coil (E_P) = 40 V
$E_P/E_S = T_P/T_S$; 40 V/E_S = 120 turns/480 turns; E_S = (40 V × 480 turns)/
120 turns; E_S = 160 V

Name _____ Date _____ Class _____

≋ SourceBook

SourceBook Assessment, continued

11. In an atom, magnetic force is caused by
a. the interaction of protons and electrons.
b. (the motion of unpaired electrons.)
c. the number of paired electrons in the atoms.
d. electrons orbiting the nucleus.

12. Magnetic domains are _____ in an unmagnetized iron bar.
a. absent   b. (randomly arranged)
c. aligned   d. None of the above

13. Which material is naturally magnetic?
a. alnico   b. magnequench   c. (magnetite)   d. iron

14. Electromagnets are _____ magnets.
a. (temporary)   b. permanent

15. When using a compass, you are working with two magnets. What two magnets are they?

The compass needle and the Earth

16. Which is *not* involved in the production of alternating current?
a. brush   b. (commutator)   c. armature   d. magnetic field

17. A transformer with two more turns in the secondary coil than in the primary coil
a. (is a step-up transformer.)   b. will decrease voltage.
c. cannot produce current.   d. Both a and c

18. A motor can be thought of as the reverse of a generator.
a. (true)   b. false

19. Match each property on the left with the correct type of force on the right. You may need to use an answer more than once.
__c__ works at a distance   a. electric
__d__ becomes stronger as   b. magnetic
distance between involved
objects increases
__c__ related to electrons   c. both a and b
__a__ determined by amount   d. neither a nor b
of charge